ISBN: 978-1-7730962-3-2

Subject Heading: MILITARY \ PTSD \ VETERANS \ MARRIAGE \ DOMESTIC RELATIONS

Printed in the United States of America

Powered by SOF Missions and The Warrior's Journey

A LETTER FROM SOF MISSIONS PRESIDENT, DR. DAMON FRIEDMAN

The ETHOS Discussion Guide is a collaborative effort between The Warrior's Journey (TWJ) and SOF Missions. TWJ is an extraordinary resource as they aim to equip veterans, families, and leaders. I want to thank TWJ's CEO, **Kevin Weaver** for his incredible devotion to helping our warriors. I also want to thank my teammate and lead author, **Dr. Dean Bonura** who was instrumental in crafting the 12 chapters of curriculum. Lastly, I want to thank **Shelley Botts** for the amazing book layout and design. It has been my honor to work this project and contribute inspiration.

F. Damon Friedman

CONTENTS

MEET THE TEAM

DAMON FRIEDMAN, D. IS.

SPECIAL TACTICS OFFICER

EDITOR, AUTHOR

Damon Friedman is an elite Special Tactics Officer and leader in special operations with four combat tours in Iraq and Afghanistan. He is the recipient of 3 Bronze Stars (one with valor) and the Combat Action medal among other awards. In 2011 He started SOF Missions, a non-profit organization aimed at helping veterans. He is the executive producer of the film SURRENDER ONLY TO ONE which creates awareness of the PTSD and suicide epidemic among the veteran community. He holds undergraduate and graduate degrees from Lewis University, University of Oklahoma, and a doctorate in Intercultural Studies from Fuller Theological Seminary. He, his wife and two kids live in Tampa, Florida.

DEAN BONURA, D. MIN.

CHAPLAIN (COL), U.S. ARMY (RET.)

LEAD AUTHOR

Dean Bonura is a retired Army chaplain with over 30 years of military service including two combat tours and several deployments to the Middle East and Bosnia-Herzegovina. He has ministered to Service Members and their families at military installations in the United States, Europe, and the Middle East, and has served at several levels of command from battalion to installation. He is the recipient of the Legion of Merit and the Bronze Star, among other awards. He also is the author of Beyond Trauma: Hope and Healing for Warriors (2016) and holds undergraduate and graduate degrees from Corban University, Western Seminary, and a doctorate from Gordon-Conwell Theological Seminary. He currently serves as a lead writer and is on the Board of Advisors for The Warrior's Journey. He and his wife, Denise, live in Memphis, Tennessee.

INTRODUCTION

AN *ETHOS* IS A WAY OF LIFE, A SET OF CORE VALUES OR PRINCIPLES THAT SHAPE A PERSON'S OR A PEOPLE'S WORLDVIEW.

Our forefathers, those legends living during the time of the American Revolution, were hard-core men. They were the knights of that era, embodying the very essence of patriotism. They exuded leadership and set the example for all to follow. They also understood the odds as they faced the British Empire which had the most powerful military in the world. In battle terms, the colonists were no match for the British. The shocking truth is that the colonists didn't care — they knew that fighting for liberty was the right cause. The American Revolution was a bloody and gut-wrenching war, but the colonists fought with honor. The knights never submitted to fear.

On the contrary, they looked to the heavenly hosts for power. They focused on the fundamentals of war, and they endured. By some miracle, and despite numerous losses, the colonists were victorious. I believe that in addition to their faith in God, their guiding principles or core values helped them win the war. The foundation of their ethos helped shaped what the US would build its government on.

An *ethos* is a way of life, a set of core values or principles that shape a person's or a people's worldview. It represents the essence of who they are, what they aspire to become, and all they resolve to do. Core values instill a sense of identity, a higher purpose, and intrinsic value. A knight armed with a solid ethos will have the proper fighting spirit -- a spirit that affirms life over death, liberty over tyranny, and justice over oppression. In Numbers 14:9, Joshua is looking out over the people of Israel as they are about to engage in battle and says, "The Lord is with us! Don't be afraid." I believe God has called you to be a knight, to engage in His righteous fight, a fight against evil in this world. The enemy is seeking to destroy. God's desire is for His knights to persevere. You must possess a robust set of principles and exercise them with vigilance. These principles will guide you in war and help you choose what's right during dark times.

Like our forefathers before us, we are fighting to liberate the oppressed. People are lost and hurting. They are looking for hope. As I look around today, I ask myself, "Where are the knights?" Today, America is one of the freest countries in the world — few nations have ever known such liberties. Unfortunately for so many today, the ethos held by our forefathers has gone by the wayside, forgotten and not lived out in our daily lives. Every knight knows that you have to be properly prepared to win. A knight has courage and is ready to defend a nation — it's his duty. You must train hard and have discipline. If you ask me, we need more knights today! We need knights who will stand by their guiding principles and fight.

- Will you fight for your *faith*?
- Will you fight for your *family*?
- Will you fight for your *friends*?
- Will you fight to the *finish*?

These are serious questions we must all answer. The enemy may seem overwhelming and impossible to defeat, but he can be beaten. How do I know this? Because with God, all things are possible! Have faith in the one who has called you and He will give you strength to face what comes before you.

Our goal through this study is that you will gain increased insight and a clearer perspective on how each core value equips your overall spiritual fitness in the Kingdom. I pray you use this study to its fullest as you strengthen your values and be the knight God has called you to be.

HAVE FAITH IN THE ONE WHO HAS CALLED YOU, AND HE WILL GIVE YOU STRENGTH TO FACE WHAT COMES BEFORE YOU.

DAMON FRIEDMAN

HOW TO USE THIS GUIDE

The ETHOS Discussion Guide is designed to prompt and generate discussion around core values relevant to the spiritual needs of service members, veterans, and their families.

Providing relevant Scriptures and probing questions, the guide encourages group interaction and individual reflection to promote understanding and spiritual growth.

THIS GUIDE CONSISTS OF TWELVE GROUP SESSIONS THAT ADDRESS A SPECIFIC CORE VALUE.

EACH SESSION PROVIDES FOR GROUP INTERACTION AND INDIVIDUAL WORK.

THE TWELVE SESSIONS CONTAIN FIVE SUBSECTIONS:

1. SESSION OVERVIEW
2. KEY SCRIPTURES
3. GROUP DISCUSSION
4. AFTER-ACTION REVIEW
5. PERSONAL REFLECTION

SESSION OVERVIEW:

Each session begins with a session overview that's designed for use in a small group setting. It introduces the group to the core value that will be the focus of the discussion and later individual work. Each core value is introduced with a related article. Group members should familiarize themselves with the related article before they engage in the group discussion. The group leader reads through the session overview and answers any questions from the group. As part of the session overview, the leader also introduces the growth objectives that are listed for that particular session. The objectives are tailored to the session's core value. The leader responds to any questions related to the session objectives. In subsequent sessions, the leader has the option to conduct a brief review of the previous session and allow for sharing of personal reflections from the previous session.

KEY SCRIPTURES:

Relevant Scriptures are listed for each session and are used to enhance group discussion and aid in individual work that's conducted between sessions. Group members may read each Scripture out loud or silently, and leaders may clarify questions that arise from the readings. Members of the group will have opportunities later in other sections to respond to questions related to Key Scriptures.

GROUP DISCUSSION:

Group discussion focuses on the articles. They form the basis for the discussion questions and individual work. Leaders may have group members read the articles privately or in smaller teams of three or four people depending on the size of the larger group. Space is provided in the guide for members to record their own responses to the questions as well as responses from others. Leaders may also have teams work on discussion questions separately, reporting back to the larger group after they've answered the assigned discussion questions.

AFTER ACTION REVIEW:

This section is designed to summarize the group session, capturing the major lessons learned, and way ahead for members of the group. It is intended to tie the elements of the session together by integrating the Scriptures, and the related articles. The leader may ask several cognitive, emotive, and behavioral questions such as: "What did you learn from this session?" "What emotions did you experience?" Or, "What connections did you make to the article?" And, "How will this session help you in your faith journey?" Or, "What actions will you take from here?" The section is designed to bring closure to the session's topic and ensure the group is prepared to do their individual work and be ready for their next session. Leaders also ensure group members feel safe and comfortable before they leave the meeting.

PERSONAL REFLECTION:

Each member of the group is given an opportunity to further reflect on the material covered in the group session. The section is completed individually after the group session and focused on personalizing the material. It includes an optional "Let's Go Deeper" portion that's geared to more interaction with the Scriptures. During the Session Overview for subsequent sessions, the leader may ask members of the group to voluntarily share their reflections from the previous session. Leaders will allow a few minutes for sharing, and then take a few moments to summarize the previous session as well as introduce the new session's challenge.

SESSION ONE

Character

SESSION OVERVIEW:

We are the sum of all our thoughts. That is the essence of character. From that flows a life, and from that life, an eternity. American essayist Ralph Waldo Emerson (1803-1882) put it this way: "Sow a thought, and you reap an action; sow an act, and you reap a habit; sow a habit, and you reap a character; sow a character, and you reap a destiny."

The development and maintenance of human character are critical. In this session we'll consider many Scriptures that speak to these two things about character, reminding us that character is ultimately about our thinking: what our minds contemplate is eventually revealed in what we say, what we do, and how we live. These are the elements of character. Chaplain Causey will tell us that good character is about using the right material—thinking about the right things or sowing with the right seed.

Like the headwaters of a mighty river, the formation of character originates in the deep and still springs of moral thought; from there flowing forth in many actions and their corresponding effects creating good and bad habits that inevitably create the warrior and determine our destiny.

GROWTH OBJECTIVES

- TO CONSIDER THE IMPORTANCE OF DEVELOPING GOOD CHARACTER.
- TO DISCERN THE FIBERS OF CHARACTER, RESULTING IN BETTER OUTCOMES.
- TO EXAMINE PATTERNS OF THINKING AND DEVELOPMENT OF HABITS AND BEHAVIORS.
- TO ESTABLISH NEW PATTERNS OF THINKING AND HEALTHY HABITS.

KEY SCRIPTURES

PSALM 1:1-2

Blessed is the man who walks not in the counsel of the wicked, nor stands in the way of sinners, nor sits in the seat of scoffers, but his delight is in the law of the LORD, and on his law, he meditates day and night.

PROVERBS 4:23

Keep your heart with all vigilance, for from it flow the springs of life.

GALATIONS 6:7-10

Do not be deceived: God is not mocked, for whatever one sows, that will he also reap. For the one who sows to his flesh will from the flesh reap corruption, but the one who sows to the Spirit will from the Spirit reap eternal life. And let us not grow weary of doing good, for in due season we will reap, if we do not give up. So then, as we have opportunity, let us do good to everyone, and especially to those who are of the household of faith.

PROVERBS 9:10

The fear of the LORD is the beginning of wisdom, and the knowledge of the Holy One is insight.

ROMANS 5:1-5

Therefore, since we have been justified by faith, we have peace with God through our Lord Jesus Christ. Through him, we have also obtained access by faith into this grace in which we stand, and we rejoice in hope of the glory of God. More than that, we rejoice in our sufferings, knowing that suffering produces endurance, and endurance produces character, and character produces hope, and hope does not put us to shame, because God's love has been poured into our hearts through the Holy Spirit who has been given to us.

ROMANS 12:1-2

I appeal to you, therefore, brothers, by the mercies of God, to present your bodies as a living sacrifice, holy and acceptable to God, which is your spiritual worship. Do not be conformed to this world, but be transformed by the renewal of your mind, that by testing you may discern what is the will of God, what is good and acceptable and perfect.

ECCLESIASTES 12:13

The end of the matter; all has been heard. Fear God and keep his commandments, for this is the whole duty of man.

EPHESIANS 5:3; 8b-10

But sexual immorality and all impurity or covetousness must not even be named among you, as is proper among saints.

Let there be no filthiness nor foolish talk nor crude joking, which are out of place, but instead let there be thanksgiving. Walk as children of light (for the fruit of light is found in all that is good and right and true), and try to discern what is pleasing to the Lord.

COLOSSIANS 3:12-14

Put on then, as God's chosen ones, holy and beloved, compassionate hearts, kindness, humility, meekness, and patience, bearing with one another and, if one has a complaint against another, forgiving each other; as the Lord has forgiven you, so you also must forgive. And above all these put on love, which binds everything together in perfect harmony.

2 TIMOTHY 2:15

Do your best to present yourself to God as one approved, a worker who has no need to be ashamed, rightly handling the word of truth.

COLOSSIANS 4:6

Let your speech always be gracious, seasoned with salt, so that you may know how you ought to answer each person.

1 TIMOTHY 6:11-12

But as for you, O man of God, flee these things. Pursue righteousness, godliness, faith, love, steadfastness, gentleness. Fight the good fight of the faith. Take hold of the eternal life to which you were called and about which you made the good confession in the presence of many witnesses.

PHILIPPIANS 2:3-4

Do nothing from rivalry or conceit, but in humility count others more significant than yourselves. Let each of you look not only to his own interests but also to the interests of others.

PHILIPPIANS 4:8-9

FINALLY, BROTHERS, WHATEVER IS TRUE, WHATEVER IS HONORABLE, WHATEVER IS JUST, WHATEVER IS PURE, WHATEVER IS LOVELY, WHATEVER IS COMMENDABLE, IF THERE IS ANY EXCELLENCE, IF THERE IS ANYTHING WORTHY OF PRAISE, THINK ABOUT THESE THINGS. WHAT YOU HAVE LEARNED AND RECEIVED AND HEARD AND SEEN IN ME—PRACTICE THESE THINGS, AND THE GOD OF PEACE WILL BE WITH YOU.

CHARACTER: ONLY AS GOOD AS THE MATERIAL IT'S MADE OF

CHAPLAIN (LIEUTENANT COLONEL) DAVID CAUSEY, U.S. ARMY (RET)

KEEP YOUR HEART WITH ALL DILIGENCE, FOR OUT OF IT SPRING THE ISSUES OF LIFE. PROVERBS 4:23

Without it, we would have never built the pyramids, crossed the oceans, or scaled Mount Everest. It has saved countless lives, pulled thousands from the angry seas, and stopped the fall of many mountain climbers. What is it?

Rope.

Rope has been around for a long time. Throughout its history—from the early Egyptians, through the Industrial Age, down to the present—rope has taken on the same basic pattern. This pattern consists of fibers twisted (clockwise) into yarns, yarns twisted (counterclockwise) into strands, and strands (clockwise again) into rope. This alternation of the twist is the secret that gives rope its cohesion and strength.

The material for rope has also remained unchanged. Earliest mankind used sinew, but people have always used plant fibers. Today the tough fibers from the bark or leaves of abaca, sisal, jute, or hemp provide the best organic material. The best modern synthetic material is nylon, polypropylene, or Dacron. Far inferior—more easily obtainable—material has also been used for rope. A rope is only as good as the fibers that go into it.

In a way, the quality of fibers that make a rope parallels the quality of the thoughts, words, deeds, and habits that form our character. Character determines our destiny.

The example of an English manufacturer of rope graphically illustrates this idea. The man lived in the early days of the Industrial Age. His arduous work and attention to detail earned him the reputation for producing the finest and most durable rope in the world. But the man began to pay more attention to profit than to quality. He sought out cheaper materials for fibers and substituted the superior fibers he had been using for inferior ones. He amassed a fortune in the process and managed to keep his reputation unstained.

Then one day, as fate would have it while sailing the Atlantic to America, a tempest-driven wave washed the man overboard. The sailors scrambled to rescue him. They tossed him a rope, which he managed to grasp. But as the sailors pulled him through the buffeting waves, the rope suddenly snapped. The man drifted hopelessly away from the ship to his destiny. Upon inspection, the seamen observed that the rope was not only new—it was the famous rope made by the very man they had sought to rescue. He had died a victim of his inferior rope. The inferior quality of his rope, like the quality of his character, had altered his destiny.

The Scripture says, "Keep your heart with all diligence, for out of it spring the issues of life." (Proverbs 4:23, NKJV)

DISCUSSION QUESTIONS

THE FOLLOWING DISCUSSION QUESTIONS ARE BASED ON THE SESSION OVERVIEW, THE ARTICLE, "CHARACTER: ONLY AS GOOD AS THE MATERIAL IT'S MADE OF," AND KEY SCRIPTURES FOR THIS SESSION.

1. Proverbs 4:23 instructs the reader to "Keep your heart with all diligence." The "heart" in Scripture often refers to the mind, and this is the reference being made here. How can you protect your mind? Why is this so important?

2. In David's article, he says the rope failed because of inferior material. What kind of materials would you use to form quality habits? How would those habits build strong character?

3. Inferior fibers, while producing greater profit for the rope maker, created inferior rope, and at a critical moment, led to a tragic end for the rope manufacturer. How does the law of sowing and reaping apply in this situation? How might it apply to yours?

4. If character determines destiny, then what changes do you need to make today so that your life results in the future you desire? What Key Scriptures might help you fulfill that destiny?

5. Character is often formed on the anvil of adversity. What challenges are you currently facing? How are those challenges affecting the development of your character?

AFTER ACTION REVIEW

1. What new information did you learn about developing character?

2. What are some of the ways a warrior can establish new patterns of thinking that form quality actions, resulting in good habits and character?

3. If character determines destiny, then where are you heading? Do you need to change azimuth? How so?

4. How has this session helped you in your spiritual journey?

PERSONAL REFLECTION

1. Why is it critical for the warrior to understand what has been called "The law of sowing and reaping"? What are some lessons you might glean from Galatians 6:7-10?

2. We learned in this session that ropes are made of quality or inferior fibers. What are some quality fibers in your life or which ones would you include to build a strong character? Which ones would you discard?

3. How has this session helped you grow in your faith or your understanding of what God expects of you?

LET'S GO DEEPER

REVIEW PSALM 1:1-2.

1. Complete the following verse by filling in the blanks:

   Blessed is the man who __________ not in the counsel of the wicked, nor __________ in the way of sinners, nor __________ in the seat of scoffers; but his delight is in the law of the LORD, and on his law he __________ day and night.

2. There is a clear contrast depicted in these verses: the way of the wicked versus the way of the godly man. According to this passage, what is the focus of the godly man? What does he think about?

3. What does the passage say about the godly man?

4. What does this passage infer about the company you keep?

REVIEW ROMANS 12:1-2.

1. What are the three actions commanded in this text?

   To __________ my body as a living sacrifice to God.

   To not be __________ to this world.

   To be __________ by the renewal of my mind.

2. In your own words, what do these commands mean to you? What do they imply about character?

LET'S GO DEEPER CONT'D:

REVIEW PHILIPPIANS 4:8-9.

1. What should warriors of character think about?

2. What do you think the writer means when he says, "What you have learned and received and heard and seen in me—practice these things, and the God of peace will be with you"?

3. How does practice contribute to character?

NOTES

SOW A THOUGHT, AND YOU REAP AN ACTION;
SOW AN ACT, AND YOU REAP A HABIT;
SOW A HABIT, AND YOU REAP A CHARACTER;
SOW A CHARACTER, AND YOU REAP A DESTINY.

RALPH WALDO EMERSON

SESSION TWO

Discipline

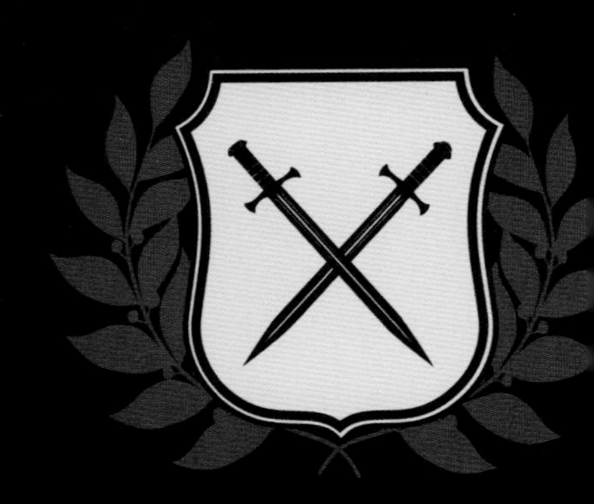

SESSION OVERVIEW:

Discipline is critical to the "warrior's kit bag." Warriors, units, and armies are only as effective as their discipline. Discipline leads to victory; indiscipline means defeat. The U.S. Army defines discipline "at the individual level [as]... the ability to control one's behavior...willingly doing what is right. Discipline is a mindset for a unit or an organization to practice sustained, systematic actions to reach and sustain a capability to perform its military function" (see ADRP 6-22, Army Leadership). All the military services define discipline similarly.

In this session, we'll explore boundaries and consider ways to achieve a balance between work and family. While that is only one aspect of discipline, it is a crucial issue for warriors who can quickly lose their balance between work and family and suffer from its effects.

GROWTH OBJECTIVES

- TO RECOGNIZE THE DANGERS OF MISPLACED PRIORITIES AND INAPPROPRIATE BOUNDARIES.
- TO UNDERSTAND GOD'S EXPECTATIONS ABOUT LIVING A DISCIPLINED LIFESTYLE.
- TO ESTABLISH USEFUL PATTERNS OF BEHAVIOR THAT CONTRIBUTE TO THE BALANCE BETWEEN WORK AND FAMILY.

KEY SCRIPTURES

PROVERBS 4:26; 5:21-23

Ponder the path of your feet; then all your ways will be sure.

For a man's ways are before the eyes of the LORD, and he ponders all his paths. The iniquities of the wicked ensnare him, and he is held fast in the cords of his sin. He dies for lack of discipline, and because of great folly he is led astray.

MATTHEW 6:19-21

Do not lay up for yourselves treasures on earth, where moth and rust destroy and where thieves break in and steal, but lay up for yourselves treasures in heaven, where neither moth nor rust destroys and where thieves do not break in and steal. For where your treasure is, there your heart will be also.

PHILIPPIANS 2:16

Holding fast to the word of life, so that in the day of Christ I may be proud that I did not run in vain or labor in vain.

MATTHEW 6:33-34

But seek first the kingdom of God and his righteousness, and all these things will be added to you. Therefore, do not be anxious about tomorrow, for tomorrow will be anxious for itself. Sufficient for the day is its own trouble.

EPHESIANS 5:8b-11, 15-17

Walk as children of light (for the fruit of light is found in all that is good and right and true) and try to discern what is pleasing to the Lord. Take no part in the unfruitful works of darkness, but instead expose them.

Look carefully then how you walk, not as unwise but as wise, making the best use of time, because the days are evil. Therefore, do not be foolish, but understand what the will of the Lord is.

MATTHEW 6:24

No one can serve two masters, for either he will hate the one and love the other, or he will be devoted to the one and despise the other. You cannot serve God and money.

KEY SCRIPTURES CONT'D:

1 TIMOTHY 4:7b-8

Rather train yourself for godliness; for while bodily training is of some value, godliness is of value in every way, as it holds promise for the present life and also for the life to come.

ROMANS 12:1-2

I appeal to you therefore, brothers, by the mercies of God, to present your bodies as a living sacrifice, holy and acceptable to God, which is your spiritual worship. Do not be conformed to this world, but be transformed by the renewal of your mind, that by testing you may discern what is the will of God, what is good and acceptable and perfect.

1 CORINTHIANS 9:24, 26-27

Do you not know that in a race all the runners run, but only one receives the prize? So, run that you may obtain it.

So, I do not run aimlessly; I do not box as one beating the air. But I discipline my body and keep it under control, lest after preaching to others I myself should be disqualified.

COLOSSIANS 3:1-2

IF THEN YOU HAVE BEEN RAISED WITH CHRIST, SEEK THE THINGS THAT ARE ABOVE WHERE CHRIST IS, SEATED AT THE RIGHT HAND OF GOD. SET YOUR MINDS ON THE THINGS THAT ARE ABOVE, NOT ON THINGS THAT ARE ON EARTH.

GAIN AND MAINTAIN MARGINS

LT. COL. DANNY WHITE II, USMC (RET)

TO REALIZE BALANCE, YOU CAN'T FOCUS JUST ON WORK.

During my formative years as a young Marine, I observed many leaders who were workaholics. Since they were successful, I assumed that I must do the same to be recognized with promotions and medals.

INVESTMENT AND BALANCE

During my journey, I had a few—and I emphasize few—leaders who cared about me as an individual and challenged me to gain and maintain margins in my life. Unfortunately, I did not understand the importance of this principle.

When looking down in the caskets of my pregnant wife Jenny and five-year-old son Danny, killed during a single-vehicle accident during a 1997 military move, I had waves of regret wash over me. I regretted those long days at work when I was working mostly to impress others. I could not hit "Rewind" then "Play" again on those days of working late at the office. I'm not advocating shirking your duties in this quest for gaining and maintaining balance. What I mean is after completing your duties for the day, avoid the temptation to remain at work for a "good idea" project, then another and another—doing this again and again.

CIRCLE OF DEPLETION

As a result of this excessive focus on work, you will fail to have margins in your life and may find yourself experiencing burn out. This point was reinforced during my assignments at the Pentagon. I saw other servicemembers working 15 to 16 hours per day, week after week. Then, when a crisis would occur in the world, these same officers didn't have any capacity to surge for the long hours of planning for the disaster and then standing watch for eight to twelve hours.

A way to combat being a workaholic during the time of "normal" operations is to establish margins in your life. This often means learning to say the powerful word "No." With only 24-hours per day or 168-hours per week, you can't do everything. To realize balance, you can't focus just on work. You must consider having some downtime, perhaps taking up a hobby or continuing one from years past.

BOUNDARIES V. LAZINESS

One story that helped me grasp a picture of margins was one my dad told about a man he admired and who helped teach him how to use a chainsaw safely. My dad would remind my brother Jon and me, while we were helping him cut dead trees for firewood to heat our home, of this story when our youthful zeal would overwhelm our better judgment to work steadily. This man, Billy Styles, would accept woodcutting jobs from various people in his rural Mountain View community. Each time before beginning his woodcutting job for a homeowner, Styles would explain that he could cut more wood if the client would allow him to stop every 20 minutes to refuel his chainsaw and sharpen the chain.

The wise homeowners would allow Styles to take this break. And Styles would steadily work between breaks and cut stacks and stacks of wood. However, the foolish homeowners would accuse Styles of being lazy and not allow him to stop except to refuel his chainsaw. As the chain became more and more dull, Styles was able to cut less and less wood. He had to work harder and harder with the dull chain to cut the wood. These homeowners failed to realize that this so-called "laziness" was a help to their bottom line. Styles still got paid the same rate per hour. The foolish homeowners received less firewood, all the while believing they had prevented Styles from being lazy.

REFLECTION

So have you considered analyzing your life for balance? Have you taken the time to sharpen your proverbial chain? If not, there may be other impacts on your journey.

MISSION V. RESPONSIBILITY

Another point to remember is that even though you are a crucial member of your organization, you are not indispensable. If you possibly became sick for an extended period, your team very likely would discover how to get things done in your absence.

After Jenny and Danny's deaths, the Marine Corps reassigned me to an Inspector-Instructor staff in Greenville, South Carolina. This assignment allowed me to assess my new role as a single parent while exploring the feasibility of still serving as a combat engineer. For four months, I visited various duty stations and met with leaders for advice on whether to resign my commission and get out of the Marine Corps or continue serving. It was an incredible slap of reality that the Marine Corps didn't stop functioning without me serving as part of a combat engineer unit. The Corps didn't grind to a halt. In fact, the Corps continued to march along. They did not even miss a beat with me not on deck. I wish that I had written down this lesson learned and reviewed it consistently after deciding to continue serving on active duty.

ALIGNMENT AND PRIORITIES

This lesson-learned aligns with an observation I made as a brand-new Marine completing my initial training at Quantico, Virginia. Our training class received a task, as part of our drill and ceremonies training, to put on a sunset parade for retiring Marines. They selected me to serve as a member of the parade staff. This provided the opportunity to stand out in front of the formation in the center of the parade deck. From this position, I could see all the retiring Marines and their families, and the medals, certificates, and flags presented to them.

That beautiful August evening scene struck me and became forever burned into my memory. I remember a Marine standing by himself when his turn came to be recognized. (For the record, this Marine may have been single his entire twenty-plus year career.) They read his retirement orders and retirement award citation. Then, the senior officer pinned the medal on his chest and handed him an American flag. After this, the Marine stepped off smartly and returned to his assigned seat. I remember thinking, "That would be so lonely to be at your retirement ceremony (after at least 20 years of service) and be standing there by yourself. I want my family to be there when I retire."

HUGS OR BRASS

What I failed to remember was to write down this important lesson-learned/observation. After joining my first active duty unit, I slowly turned into a workaholic. Then years later as a senior officer, I almost destroyed my marriage due to being a workaholic at the Pentagon. If I had not had a close wake-up call, I likely would have received another promotion and another medal. This would have resulted in me standing alone at my retirement ceremony—due to losing my family in the process.

I remarried after losing my wife in the vehicle accident. Years later when I retired, it was a joy to me knowing my family was by my side. That experience far outweighed any promotion or award.

While the medals and rank insignia are packed away in a box, the relationships with my wife and children continue to grow stronger and stronger, for which I'm eternally grateful.

DISCUSSION QUESTIONS

THE FOLLOWING DISCUSSION QUESTIONS ARE BASED ON THE SESSION OVERVIEW, THE ARTICLE, "GAIN AND MAINTAIN MARGINS," AND KEY SCRIPTURES FOR THIS SESSION.

1. It took a tragedy for Lieutenant Colonel White to realize he had become a workaholic. In what areas of work and family are you vulnerable to similar imbalance? What can you do today to begin to correct that imbalance?

2. How would you define inappropriate boundaries in the workplace? What are the dangers? In what other areas are warriors susceptible to creating inappropriate boundaries or misplaced priorities?

3. Identify some practical ways a warrior might establish and maintain appropriate margins (or boundaries) between life and work. Given the profession of arms and the demands placed on warriors, what are the challenges in fulfilling the need for appropriate margins between life and work?

4. Only two things are eternal: God's Word and people. How does this truth inform your priorities for living?

5. What can leaders do to assist subordinates in establishing healthy priorities and proper balance between work and family?

AFTER ACTION REVIEW

1. What new information did you learn about establishing priorities and achieving a balance between work and family?

2. In what other areas does God expect you to achieve balance? How does this reflect discipline?

3. How has this session helped you in your spiritual journey?

PERSONAL REFLECTION

1. What actions might you take based on this session?

2. Identify two or three areas in your life that require your re-evaluation.

LET'S GO DEEPER

REVIEW MATTHEW 6:19-21, 24, 33-34.

1. What does Matthew 6:19-20 say about treasures?

2. What do these verses say about priorities?

NOTES

HAVE THE DISCIPLINE TO FOCUS YOUR TIME AROUND YOUR PRIORITIES. THE MOST MEANINGFUL THINGS IN YOUR LIFE SHOULD NEVER BE SACRIFICED TO THOSE THAT ARE THE LEAST MEANINGFUL.

ROBIN SHARMA

SESSION THREE

Courage

SESSION OVERVIEW:

Courage consists of mental and moral strength to speak or act when it is the right thing to do, and usually at times of significant personal risk and cost. Courage operates in the face of fear; but never succumbs to fear. The biblical idea of courage always has the quality of moral strength; but also includes concepts of vigilance, firmness, and determination. In this session, you'll consider Key Scriptures that illustrate or commend moral courage and review an article titled "When You Feel Overwhelmed."

The article refers to the fear of failure, and how warriors can face and overcome their fear. Courage is not the quality or capability that helps a person do or say what is right without fear, but the strength to do what is necessary regardless of fear.

GROWTH OBJECTIVES

- TO EXAMINE THE MEANING OF COURAGE.
- TO IDENTIFY OBSTACLES THAT THWART COURAGE.
- TO IDENTIFY THE QUALITIES OF COURAGE.
- TO EXPLORE WAYS TO APPLY THE QUALITIES OF COURAGE.

KEY SCRIPTURES

JOSHUA 1:5-7, 9

No man shall be able to stand before you all the days of your life. Just as I was with Moses, so I will be with you. I will not leave you or forsake you. Be strong and courageous, for you shall cause this people to inherit the land that I swore to their fathers to give them. Only be strong and very courageous, being careful to do according to all the law that Moses my servant commanded you. Do not turn from it to the right hand or to the left, that you may have good success wherever you go.

Have I not commanded you? Be strong and courageous. Do not be frightened, and do not be dismayed, for the LORD your God is with you wherever you go.

NUMBERS 13:27-28a, 30-31

And they told him, 'We came to the land to which you sent us. It flows with milk and honey, and this is its fruit. However, the people who dwell in the land are strong, and the cities are fortified and very large.'

But Caleb quieted the people before Moses and said, 'Let us go up at once and occupy it, for we are well able to overcome it.' Then the men who had gone up with him said, 'We are not able to go up against the people, for they are stronger than we are.'

JOSHUA 10:25a

And Joshua said to them, 'Do not be afraid or dismayed; be strong and courageous.'

2 CHRONICLES 15:2, 7-8

And [Azariah, the prophet] said to [Asa, the king] 'Hear me, Asa, and all Judah and Benjamin: The LORD is with you while you are with him. If you seek him, he will be found by you, but if you forsake him, he will forsake you.

But you, take courage! Do not let your hands be weak, for your work shall be rewarded. As soon as Asa heard these words...he took courage and put away the detestable idols from all the land...and repaired the altar of the LORD.

PSALM 27:1, 14

The LORD is my light and my salvation; whom shall I fear? The LORD is the stronghold of my life; of whom shall I be afraid?

Wait for the LORD; be strong, and let your heart take courage; wait for the LORD!

DANIEL 3:16-18

Shadrach, Meshach, and Abednego answered and said to the king, O Nebuchadnezzar, we have no need to answer you in this matter. If this be so, our God whom we serve is able to deliver us from the burning fiery furnace, and he will deliver us out of your hand, O king. But if not, be it known to you, O king, that we will not serve your gods or worship the golden image that you have set up.

1 CORINTHIANS 16:13

Be watchful, stand firm in the faith, act like men, be strong.

2 TIMOTHY 1:6-7

For this reason, I remind you to fan into flame the gift of God, which is in you through the laying on of my hands, for God gave us a spirit not of fear but of power and love and self-control.

2 TIMOTHY 1:12b

But I am not ashamed, for I know whom I have believed, and I am convinced that he is able to guard until that Day what has been entrusted to me.

ISAIAH 41:10

FEAR NOT, FOR I AM WITH YOU; BE NOT DISMAYED, FOR I AM YOUR GOD; I WILL STRENGTHEN YOU, I WILL HELP YOU, I WILL UPHOLD YOU WITH MY RIGHTEOUS RIGHT HAND.

WHEN YOU FEEL OVERWHELMED

CHAPLAIN (LIEUTENANT COLONEL) DAVID CAUSEY, U.S. ARMY (RET)

JUST AS I HAVE BEEN WITH MOSES, I WILL ALSO BE WITH YOU; I WILL NEVER LEAVE YOU OR FORSAKE YOU. ONLY BE STRONG AND COURAGEOUS. *JOSHUA 1:5-6*

Have you ever felt like you're in over your head? Ever feel like a task is so large that you cannot even begin? If so, remember these simple thoughts:

THE LONGEST JOURNEY BEGINS WITH A SINGLE STEP.

The most significant accomplishments all begin with a dream and the courage to try to make it a reality. Most people never start because they fear failure. But it is better to try and do something imperfectly than to do nothing flawlessly. As Joshua stood at the threshold of the Promised Land, realizing the enormous and hazardous task that lay before him, God told him: "Just as I have been with Moses, I will also be with you; I will never leave you or forsake you. Only be strong and courageous" (Joshua 1:5-6).

BEGINNING IS HALFWAY THERE.

Just getting started is half the battle. Once you get started, you will gain momentum. Pastor and author, Robert Schuller said, "The one battle most people lose is the battle over the fear of failure...Try, start, begin and you'll be assured you won the first round!" Many others, with less talent and ability than you, have come this way before and have made it. So can you. Get started!

INCH-BY-INCH, ANYTHING'S A CINCH!

God told Joshua he did not have to conquer the Promised Land all in one year, but "little by little" (Exodus 23:29-30). So, God also commands us to take just one day at a time, to only bear today's burdens and to place tomorrow's concerns in His almighty hands (Matthew 6:34; Psalm 55:22).

GOD PROMISES TO GIVE YOU ONLY WHAT YOU CAN BEAR!

In 1 Corinthians 10:13, God tells us that He only allows trials to enter our lives that everybody else must face and that He will never push us beyond our limitations. Put your trust in God; He will not let you down.

THERE IS AN END TO EVERY TASK, EVERY TRIAL, EVERY TEST, AND EVERY ORDEAL. IT WILL ALL SOON BE HISTORY!

Robert Schuller states, "Every problem has a limited life-span," and "Tough times don't last, but tough people do!" More important, though our trials are not forever, God's power is. Isaiah 40:28-29 tells us that the Everlasting God never grows weary but gives strength to the weary and weak. The apostle Paul tells us that "our light affliction is but for a moment" (2 Corinthians 4:17). Hang in there, and this task or trial will soon be behind you.

AS YOU ACCOMPLISH WHAT YOU ONCE THOUGHT "IMPOSSIBLE," YOU WILL MOVE FROM WEAKNESS TO STRENGTH AND BE TRANSFORMED.

Phillips Brooks, a nineteenth-century American clergyman, once said, "Do not pray for easier lives, pray to be stronger men. Do not ask for tasks equal to your powers, but powers equal to your task. Then not only shall the doing of your work be a miracle, but you shall be a miracle. Every day you shall wonder at yourself at the richness of life which has come in you by the grace of God."

DISCUSSION QUESTIONS

THE FOLLOWING DISCUSSION QUESTIONS ARE BASED ON THE SESSION OVERVIEW, THE ARTICLE, "WHEN YOU FEEL OVERWHELMED," AND KEY SCRIPTURES FOR THIS SESSION.

1. If courage is the strength to speak or act at great personal risk and cost, what situations have you faced that required the need for courage? What pressures or stresses did you feel? Are you facing similar situations now? What are you doing about them?

2. The fear of failure is a common human problem. We don't know how things will turn out. Nobody has a crystal ball. What are some areas where you face potential for failure? Given what you're learning in this session, how will you proceed?

3. What does it cost for courage, in your own life or in what you've observed in others? What are the inherent risks? What are the rewards?

4. According to the Key Scriptures, what does God promise to those who will act with courage?

5. The article says that fear of failure is an obstacle that keeps many people from achieving their potential. What other barriers prevent people from achieving their potential or completing their mission? How does moral strength help someone overcome these obstacles?

AFTER ACTION REVIEW

1. What new information did you learn about courage?

2. How can a warrior overcome the fear of failure or other similar obstacles?

3. What does God say about courage?

4. How has this session helped you in your spiritual journey?

PERSONAL REFLECTION

1. Identify three qualities of courage. How might these qualities help you in your present situation or future situations?

2. Based on Key Scriptures and the article, what resources does God promise to help you live courageously?

3. How has this session helped you grow in your faith or your understanding of what God expects of you?

LET'S GO DEEPER

REVIEW JOSHUA 1:5-7, 9.

1. Circle the words "strong" and "courageous" in the passage:

   No man shall be able to stand before you all the days of your life. Just as I was with Moses, so I will be with you. I will not leave you or forsake you. Be strong and courageous, for you shall cause this people to inherit the land that I swore to their fathers to give them. Only be strong and very courageous, being careful to do according to all the law that Moses my servant commanded you. Do not turn from it to the right hand or to the left, that you may have good success wherever you go.

   Have I not commanded you? Be strong and courageous. Do not be frightened, and do not be dismayed, for the LORD your God is with you wherever you go.

2. What does God promise Joshua?

   No man shall ___________________ before you.

   Just as I was with Moses, so I will be _______________ you.

   I will not _______________ you or _______________ you.

3. What was Joshua's mission from God?

   To cause this people to ___________________ the land.

   To ___________all the law that Moses my servant commanded you.

   I will not _______________ you or _______________ you.

LET'S GO DEEPER CONT'D:

REVIEW 2 CHRONICLES 15:2, 7-8.

1. What does God promise Asa, the king, if he would seek God?

2. God tells Asa to "take courage" and "not let your hands be weak." What did Asa do?

3. Asa was obedient to God, acting boldly and courageously to remove the idols that were polluting the land of Israel and repairing the altar of God. What is God asking you to do that would require boldness and courage?

NOTES

DO NOT PRAY FOR EASIER LIVES, PRAY TO BE STRONGER MEN. DO NOT ASK FOR TASKS EQUAL TO YOUR POWERS, BUT POWERS EQUAL TO YOUR TASK. THEN NOT ONLY SHALL THE DOING OF YOUR WORK BE A MIRACLE, BUT YOU SHALL BE A MIRACLE.

PHILLIPS BROOKS

SESSION FOUR

Honor

SESSION OVERVIEW:

Honor represents the bond that exists between warriors and society that manifests in qualities of honesty, integrity, and respect. One's sense of honor flows from esteem for what is morally right, true, and correct. Honor often reflects a commitment to a moral code, a perceived quality of worthiness, and an ideal of conduct. Samuel Johnson, in *A Dictionary of the English Language* (1775), defined honor as "nobility of soul, magnanimity, and a scorn of meanness." To give honor is to pay high respect and acknowledge worth. Honor is all about dignity, fairness, and unbending principle. In the military, honor is that encompassing quality which infuses the values of the profession of arms such as loyalty, sacrifice, and commitment.

GROWTH OBJECTIVES

- TO REFLECT ON THE MEANING OF HONOR AND HOW IT APPLIES TO OUR LIVES.
- TO GAIN GREATER INSIGHT INTO THE IMPLICATIONS OF LIVING AN HONORABLE LIFE.
- TO ADOPT NEW WAYS OF THINKING AND BEHAVING.

KEY SCRIPTURES

1 PETER 2:17

Honor everyone. Love the brotherhood. Fear God. Honor the emperor.

ROMANS 12:10, 13:7

Love one another with brotherly affection. Outdo one another in showing honor.

Pay to all what is owed to them: taxes to whom taxes are owed, revenue to whom revenue is owed, respect to whom respect is owed, honor to whom honor is owed.

JOHN 5:23

That all may honor the Son, just as they honor the father. Whoever does not honor the Son does not honor the Father who sent him.

HEBREWS 13:4

Let marriage be held in honor among all, and let the marriage bed be undefiled, for God will judge the sexually immoral and adulterous.

EXODUS 20:12

Honor your father and your mother, that your days may be long in the land that the LORD your God is giving you.

LEVITICUS 19:32

You shall stand up before the gray head and honor the face of an old man, and you shall fear your God: I am the LORD.

PROVERBS 3:9-10

Honor the LORD with your wealth and with the firstfruits of all your produce; then your barns will be filled with plenty, and vats will be bursting with wine.

1 TIMOTHY 6:1

Let all who are under a yoke as slaves regard their own masters as worthy of all honor, so that the name of God and the teaching may not be reviled.

1 PETER 3:7

Likewise, husbands, live with your wives in an understanding way, showing honor to the woman as the weaker vessel, since they are heirs with you of the grace of life, so that your prayers may not be hindered.

1 SAMUEL 2:27-32

And there came a man of God to Eli and said to him, 'Thus the LORD has said, 'Did I indeed reveal myself to the house of your father when they were in Egypt subject to the house of Pharaoh? Did I choose him out of all the tribes of Israel to be my priest, to go up to my altar, to burn incense, to wear an ephod before me? I gave to the house of your father all my offerings by fire from the people of Israel. Why then do you scorn my sacrifices and my offerings that I commanded, and honor your sons above me by fattening yourselves on the choicest parts of every offering of my people Israel? Therefore the LORD, the God of Israel, declares: 'I promised that your house and the house of your father should go in and out before me forever,' but now the LORD declares: 'Far be it from me, for those who honor me I will honor, and those who despise me shall be lightly esteemed.'

PROVERBS 21:21

WHOEVER PURSUES RIGHTEOUSNESS AND KINDNESS WILL FIND LIFE, RIGHTEOUSNESS, AND HONOR.

ARE YOU AN IMPOSTER?

CHAPLAIN (LIEUTENANT COLONEL) DAVID CAUSEY, U.S. ARMY (RET)

LET EVERYONE WHO NAMES THE NAME OF THE LORD ABSTAIN FROM WICKEDNESS. *2 TIMOTHY 2:19*

Imposter. We all know what the word means. The very mention of the word probably brings someone to mind. Maybe someone who pretends to be someone better, more interesting, more significant than he or she is.

Does anyone remember Rosie Ruiz? In 1980 she was once heralded as the world record-breaking women's Boston Marathon champion—for a day or so. Then it was discovered that Rosie had hopped aboard public transportation early in the race and hopped off near the finish line. She crossed the finish line ahead of every other woman in record time. An investigation disclosed she had taken a similar shortcut in a previous New York Marathon. Besides being a cheat, Rosie pretended to be a marathon champion, relishing in all the glory she never earned. Our world abounds with such pretenders.

There have also been many outlaw impersonators. Historically, Jesse James seems to have been a favorite with imposters. One James-pretender, who hadn't done his homework very well, rolled into a Kentucky town to ply his trade—a city in which the James' Gang had robbed a bank years before. A U.S. Marshall confronted the imposter, pulled a yellowed document from his pocket and declared: "I have here a warrant for your arrest, charging you with the robbery of the Bank of Columbia and the murder of the cashier." When the imposter turned pale, the Marshall added: "However, on the chance that you may not be the real Jesse James, I'm giving you ten minutes to get out of town." The man was gone in five.

But of all imposters, the worst may have been the "Great Pretender" himself—Ferdinand Waldo Demara. Without training, academic degrees, credentials, ordination or a commission, Demara pretended to be (and received payment as) a surgeon, psychologist, college dean, dentist, university professor, naval officer, and a Trappist monk. Demara's phony life finally caught up with him and landed him an 18-month prison sentence, one of the few things in life he earned.

Maybe we should add some of our own names to the list of phonies. How many of us relish in the name "Soldier/Marine/Sailor/Airman" yet do not share that branches' values? How many wear a starched uniform and spit-shined boots, yet scheme to avoid the sacrifices, training, and hardship our duty requires? Far too many, I fear. Far too many do everything to serve and protect themselves in a profession that may require them to lay down their lives. Then there are the religious imposters among us, who name the name of Christ, yet never try to live as He commanded us. The Scripture says, "Let everyone who names the name of the Lord abstain from wickedness" (2 Timothy 2:19).

DISCUSSION QUESTIONS

THE FOLLOWING DISCUSSION QUESTIONS ARE BASED ON THE SESSION OVERVIEW, THE ARTICLE, "ARE YOU AN IMPOSTER?," AND KEY SCRIPTURES FOR THIS SESSION.

1. Given your understanding of honor as described in the Session Overview, how might a warrior conduct himself honorably? Or, what do you think it means to live honorably?

2. Another word for "imposter" is a hypocrite. Hypocrites wear masks to conceal their true identities or true motives. Their talk never matches their walk. What are the dangers of living this way? How does it impact the team?

3. How does honor infuse the values of the military profession? Where do you think the pitfalls exist? How does an honorable warrior avoid those pitfalls and remain faithful to self, true to country, and true to God?

4. According to the Key Scriptures, God expects His followers to live honorably. What does God say about honor?

5. The passage found in Romans 12:10; 13:7 says we should "outdo one another in showing honor." How could you fulfill the requirements of this command? What specific examples can you come up with?

AFTER ACTION REVIEW

1. What new information did you learn about living honorably?

2. What does God say about honor and how does He expect it to manifest in a person and an organization?

3. What are the dangers of living like an imposter? How can a warrior avoid living like a hypocrite?

4. How has this session helped you in your spiritual journey?

PERSONAL REFLECTION

1. Have you ever faked it? What led you to live hypocritically? How did you overcome it?

2. Thinking of comments about the meaning of honor in the Session Overview, what concepts or values spoke loudly to you? How might you apply them to your situation today?

3. How has this session helped you grow in your faith or your understanding of what God expects of you?

LET'S GO DEEPER.

REVIEW 1 SAMUEL 2:27-32.

1. According to this passage, what did Eli fail to do?

2. What three things did God do for Eli's house?

   He ________________ himself to the house of his father.

   He ________________ him out of all the tribes of Israel to be a priest.

   God ______________ to the house of his father all his offerings.

3. What did God initially promise to the house of Eli?

4. Given Eli's failure, what did God say about giving and receiving honor in the future?

5. How does a person forfeit God's esteem?
   "Those who ___________ me shall be lightly ________________."

6. What does it mean to "despise" God?

NOTES

THE SHORTEST AND SUREST WAY TO LIVE WITH HONOR IN THE WORLD, IS TO BE IN REALITY WHAT WE WOULD APPEAR TO BE.

SOCRATES

SESSION FIVE

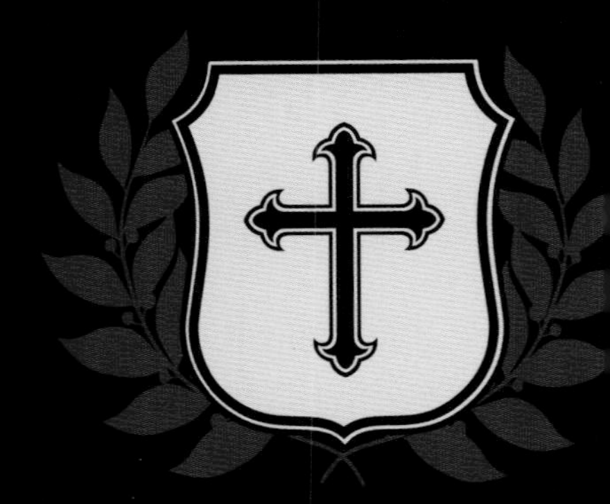

SESSION OVERVIEW:

In this session, you'll view a short video titled "Bulletproof Faith,"presented by Jeff Struecker, who recounts his experience in Somalia twenty years earlier. Jeff tells about the challenges he faced and the fears he experienced while serving in Somalia; and how he dealt with the possibility of his death. He encourages viewers to repent of their sins and turn to Christ. Only in Him will they find the faith and the strength to withstand the challenges of the battlefield.

You'll also read an article by Chaplain Dean Bonura, a fellow combat veteran, who describes the meaning of faith, and how biblical faith, faith placed in the Person of Jesus Christ, enables a warrior to have complete confidence in God. He'll tell you that everyone needs to be rightly related to God by the exercise of faith, a faith that will result in receiving the righteousness of God and the blessings of eternal life.

GROWTH OBJECTIVES

- TO UNDERSTAND THE MEANING OF BIBLICAL, "BULLETPROOF" FAITH.
- TO LEARN HOW A WARRIOR BECOMES RIGHT WITH GOD.
- TO EXPERIENCE THE TRUE MEANING OF FAITH AND FIND GOD'S PEACE.

KEY SCRIPTURES

JOHN 1:12

But to all who did receive him, who believed in his name, he gave the right to become children of God.

JOHN 5:24

Truly, truly, I say to you, whoever hears my word and believes him who sent me has eternal life. He does not come into judgment but has passed from death to life.

ROMANS 4:3

For what does the Scripture say? Abraham believed God, and it was counted to him as righteousness.

ROMANS 5:1-2

Therefore, since we have been justified by faith, we have peace with God through our Lord Jesus Christ. Through Him, we have also obtained access by faith into this grace in which we stand, and we rejoice in hope of the glory of God.

ROMANS 6:23

For the wages of sin is death, but the free gift of God is eternal life in Christ Jesus our Lord.

2 CORINTHIANS 5:21

For our sake he made him to be sin who knew no sin, so that in him we might become the righteousness of God.

PHILIPPIANS 3:9

And be found in him, not having a righteousness of my own that comes from the law, but that which comes through faith in Christ, the righteousness from God that depends on faith.

HEBREWS 11:1

NOW FAITH IS THE ASSURANCE OF THINGS HOPED FOR, THE CONVICTION OF THINGS NOT SEEN.

BULLETPROOF FAITH

CHAPLAIN (COLONEL) DEAN BONURA, U.S. ARMY (RET)

VIDEO: "BULLETPROOF FAITH" PRESENTED BY JEFF STRUECKER.
YOU CAN ACCESS THIS VIDEO AT SOFMISSIONS.COM/ETHOS

NOW FAITH IS THE ASSURANCE OF THINGS HOPED FOR, THE CONVICTION OF THINGS NOT SEEN. *HEBREWS 11:1*

How does a person become right with God? That's one of the ultimate questions of life. The answer is found in the Scriptures where God tells us that a person can be right with God through faith (Rom. 1:17; Heb. 10:38-39).

THE FAITH PERSPECTIVE

Righteousness is a gift from God. It becomes ours by faith in Christ (Phil. 3:9). So faith is pretty important. But what is faith? What is real faith, the kind of faith that puts one in a right relationship with God? Authentic faith is what Jeff Struecker calls "bulletproof faith."

Faith is only as good as its object. It's not the strength or sincerity of faith or belief, but the object of belief that determines the quality of faith.

For example, while I love my three-year-old grandson and believe he can do the things that most three-year-olds do like feed himself and get himself to the toilet, I'm not going to give him the keys to my car. He can't operate it. He can't drive. He lacks the capability. Now, one day he may ask me for the keys, but not today.

The Bible tells us in Hebrews 11:1 that "Faith is the substance of things hoped for, the evidence of things not seen" (King James Version). Another translation puts it this way, "Now faith is the assurance of things hoped for, the conviction of things not seen" (English Standard Version). There are certain future realities, "things" identified in the Bible. For example, there is the coming of Christ, the fulfillment of the Christian's inheritance, and the judgment of God. Faith does not make these things so. Faith accepts them to be so.

BIBLICAL FAITH

Biblical faith, real faith, is absolute confidence or trust in what God says and in what God does. God tells us, "faith comes by hearing and hearing by the word of Christ" (Rom. 10:17). It's about capabilities. God is capable of fulfilling every promise he makes. He's capable of achieving whatever he desires to do.

You would not have my three-year-old grandson perform brain surgery on you or anyone else. He's not capable of any surgery. Believe me. But a nationally renowned neurosurgeon that performs daily operations at the Mayo Clinic is competent. You'd trust the neurosurgeon over the three-year-old any day.

When I was in Iraq, I was given two SAPI (Small Arms Protective Insert) plates as part of my protective system (IPS). I was told the ceramic plates were capable of stopping a NATO 7.62mm round. I never had the opportunity to test that capability, but one of my soldiers did. The plate saved his life. I have confidence or trust in those plates. They work. They're bulletproof. They're the real deal. That's what Jeff is talking about—capable faith, faith that works, faith that's real. It's faith in God because God is capable of preserving our lives, capable of delivering the goods. You have His word on it!

THE OBJECT OF FAITH DETERMINES THE QUALITY OF FAITH

Faith in God and in what He says is high-quality faith; any other faith is questionable. When Jeff Struecker faced the question of whether he was going to live or die in Mogadishu, he placed his faith in God. He put his life in the hands of God. Why? He believed that whatever happens on earth, no matter what it is, God is with us. God was in charge of his life, and nothing was going to happen without God's permission.

REFLECTION

Do you have that kind of "bulletproof faith"? The type of faith that holds up under the pressures and stresses of life; that assures you of an eternity with God and forgiveness of sins? Do you have the kind of faith that makes you right with God? You can.

No one is, but you can have the "bulletproof faith": faith that's placed in the person of the Son of God—the Lord Jesus Christ.

"BULLETPROOF FAITH" DEPENDS ON THE CHARACTER AND WORK OF GOD

The Bible tells us that sin separates all of us from God. Sin prevents us from having a right relationship with God (Rom. 3:23; 6:23). There's nothing we can do to fix the sin problem on our own. You can't save yourself. You can't pay your way out of sin or work your way out of it.

We're all doomed because of sin. But God sent his Son to die for our sins. He fixed the sin problem and opened the way for us to enjoy a right relationship with God.

Jesus can free us from the condemnation brought about by our sin (John 3:18, 36). He was able to die for our sins because he never had any sin of his own (1 Pet. 2:21-22). He died for us. He was our substitute.

In a sense, like a person stuck in quicksand and doomed to perish, he rescued us from our sins. We are lost, but he can save us. He's the only one that can pull us up "*out of the miry bog*" (Ps. 40:2). That's a reality. He demonstrated his capability by rising from the dead on the third day (1 Cor. 15:3-5), and then proved it by appearing before more than five hundred people at one time. It was no apparition; it was real. He conquered death itself. And He will for you, too.

FAITH IN ACTION

So what does it take from us? It takes "bulletproof faith." You must believe that you are a sinner and that Christ died for you. You must believe that Christ died for your sins and rose again on the third day. In response to your faith, God offers his righteousness. He also offers eternal life and forgiveness of sin.

Place your complete faith in Christ today. Tell Him that you accept His sacrifice for your sins, that you accept His forgiveness of your sins, and that you will trust Him as your Lord and Savior today.

DISCUSSION QUESTIONS

THE FOLLOWING DISCUSSION QUESTIONS ARE BASED ON THE VIDEO, "BULLETPROOF FAITH," THE ARTICLE BY THE SAME TITLE, AND KEY SCRIPTURES INCLUDED IN THIS SESSION.

1. According to Jeff, what must warriors do to experience a radical change in their life?

2. Referring to the article, what is "bulletproof faith?" How does a person exercise that kind of faith? How does "bulletproof faith" change the way a person looks at life?

3. Dean says that a warrior or any person can be right with God by placing their faith in Jesus Christ. Why do we need to be right with God? Why do we need the righteousness of God?

4. Jeff says, "Jesus is real, and faith in Jesus Christ will make all the difference in your life." What did Jesus do for us? How might your faith in Jesus make all the difference for you?

5. What are the consequences of rejecting Jesus Christ? What are the benefits of believing in Jesus Christ?

AFTER ACTION REVIEW

1. What new information did you learn about the kind of faith that is "bulletproof?"

2. What must a warrior do to exercise faith in Jesus Christ?

3. What must a warrior accept about themselves and Jesus Christ to exercise "bulletproof faith?"

4. How has this session helped you in your spiritual journey?

5. What actions might you take based on this session?

PERSONAL REFLECTION

1. If you do not have a "bulletproof faith," what are the things that are keeping you from having that kind of faith?

2. Referring to Jeff's comments in the video, some of the guys from the battle of Mogadishu wondered if they'd survive the battle; they also had questions about placing their faith in God. What are your questions or concerns? Do you worry about the future? What about your fears? How might putting your complete faith in Christ relieve you of those fears?

3. How has this session helped you grow in your faith or your understanding of what God can do for you?

LET'S GO DEEPER

REVIEW JOHN 1:12; 5:24 & ROMANS 4:3.

1. Complete the sentences below:

   John 1:12: He [God] gave the right to become ______________ of God.

   John 5:24: [He or She] has ___________________ life. He [or She] does not come into _________________, but has passed from death to life.

   Romans 4:3: It was _________________ to him as righteousness.

   In Romans 4:3, the word you added was "counted." This is an accounting term. It means God deposited his righteousness into "Abraham's account" so to speak. Abraham had no righteousness of his own. He was an unrighteous sinner, but God gave him Christ's righteousness because he believed God.

2. What do these Scriptures say about what it means to believe God?

REVIEW ROMANS 5:1-2.

1. Complete the following verse by filling in the blanks:

   Therefore, since we have been ______________ by faith, we have __________ with God through our Lord Jesus Christ. Through him, we have also obtained access by ___________ into this grace in which we stand, and we rejoice in ___________ of the glory of God.

2. To be *justified* by faith is to be *made righteous* in the sight of God. What other benefits does a believer receive from God based on these two verses?

   We have _____________ with God.

   We have also obtained ____________by faith.

   We _____________ in the hope of the glory of God.

REVIEW PHILIPPIANS 3:9 & 2 CORINTHIANS 5:21.

1. What do these verses say about righteousness?

NOTES

IT IS NOT THE STRENGTH OF YOUR FAITH, BUT THE OBJECT OF YOUR FAITH THAT ACTUALLY SAVES YOU.

TIMOTHY KELLER

SESSION SIX

Duty

SESSION OVERVIEW:

A duty is a commitment to fulfill one's obligations or assigned tasks. Such requirements may be designated or implied. They usually arise from some specified expectation and may be derived from an ethical code or law. For a warrior, duty is also based in honor, accomplished for the sake of a mission, for the benefit of others or in concert with others, regardless of personal risk or cost. General Robert E. Lee said: "Do your duty in all things. You cannot do more. You should never wish to do less."

In this session, we consider a danger inherent in failing to do one's duty — the temptation to seek revenge, to express hatred and prejudice at the cost of duty. These things place self before the team or mission. They are a distraction that takes a warrior down the wrong path and inevitably leads to failure.

GROWTH OBJECTIVES

- TO REFLECT ON THE MEANING AND APPLICATION OF DUTY.
- TO COMPREHEND THE DANGERS INHERENT IN THE EXPRESSION OF HATRED, PREJUDICE, AND REVENGE.
- TO ADOPT HEALTHY WAYS OF DEALING WITH FEELINGS OF HATRED, PREJUDICE, AND REVENGE.

KEY SCRIPTURES

LEVITICUS 19:15-18

You shall do no injustice in court. You shall not be partial to the poor or defer to the great, but in righteousness shall you judge your neighbor. You shall not go around as a slanderer among your people, and you shall not stand up against the life of your neighbor: I am the LORD. You shall not hate your brother in your heart, but you shall reason frankly with your neighbor, lest you incur sin because of him. You shall not take vengeance or bear a grudge against the sons of your people, but you shall love your neighbor as yourself: I am the LORD.

PSALMS 34:13-14

Keep your tongue from evil and your lips from speaking deceit. Turn away from evil and do good; seek peace and pursue it.

PROVERBS 24:29

Do not say, 'I will do to him as he has done to me; I will pay the man back for what he has done.

PROVERBS 10:12

Hatred stirs up strife, but love covers all offenses.

ROMANS 12:9-10, 16-19

Let love be genuine. Abhor what is evil; hold fast to what is good. Love one another with brotherly affection. Outdo one another in showing honor. Live in harmony with one another. Do not be haughty, but associate with the lowly. Never be wise in your sight. Repay no one evil for evil, but give thought to do what is honorable in sight of all. If possible, so far as it depends on you, live peaceably with all. Beloved, never avenge yourselves, but leave it to the wrath of God, for it is written, 'Vengeance is mine, I will repay, says the Lord.'

TITUS 3:2-3

To speak evil of no one, to avoid quarreling, to be gentle, and to show perfect courtesy toward all people. For we ourselves were once foolish, disobedient, led astray, slaves to various passions and pleasures, passing our days in malice and envy, hated by others and hating one another.

KEY SCRIPTURES CONT'D:

I THESSALONIANS 5:15, 22

See that no one repays anyone evil for evil, but always seek to do good to one another and to everyone.

Abstain from every form of evil.

JAMES 2:1-4

My brothers, show no partiality as you hold the faith in our Lord Jesus Christ, the Lord of glory. For if a man wearing a gold ring and fine clothing comes into our assembly, and a poor man in shabby clothing also comes in, and if you pay attention to the one who wears the fine clothing and say, 'You sit here in a good place,' while you say to the poor man, 'You stand over there,' or, 'Sit down at my feet,' have you not then made distinctions among yourselves and become judges with evil thoughts.

ROMANS 15:1-3

WE WHO ARE STRONG HAVE AN OBLIGATION TO BEAR WITH THE FAILINGS OF THE WEAK, AND NOT TO PLEASE OURSELVES. LET EACH OF US PLEASE HIS NEIGHBOR FOR HIS GOOD, TO BUILD HIM UP. FOR CHRIST DID NOT PLEASE HIMSELF, BUT AS IT IS WRITTEN, “THE REPROACHES OF THOSE WHO REPROACHED YOU FELL ON ME.”

HATRED, PREJUDICE, AND REVENGE

THE WARRIOR'S JOURNEY TEAM

I HAVE COME THAT THEY MAY HAVE LIFE AND THAT THEY MAY HAVE IT MORE ABUNDANTLY. JOHN 10:10

Our battalion, which deployed to Anbar province in support of Operation Iraqi Freedom, was primarily a support unit. Only a fraction of our personnel routinely went off base, "outside the wire."

Yet among those who did go outside, we had casualties...several "killed in action." After our first such incident I heard from others who, in the midst of their trauma, grief, guilt (as those who stayed behind) and anger, voices threats dripping with hatred and vengeance: "I know the TCN's (third country nationals) at the DFAC (dining facility) are not the ones who were behind the IEDs...but I can't help it...I just want to see them dead—to kill them myself!"

Is that any way for a person of faith to speak? Would a Christian react so violently? We may have thought, even said (when we were kids, at least): "*I wish you were dead!*" but haven't we outgrown such words and aims by the time we're old enough to vote and join the military? Does our tradition, the Scriptures, the teaching of Jesus Christ have anything to say about hatred, prejudice, and revenge?

You may have heard it expressed that the dramatic, troubling, extreme reactions that come with any trauma—certainly including those related to combat operations and casualties—are "normal reactions to abnormal circumstance."

Acknowledging that the use of lethal force, up to and including military operations, may be justifiable, we are also right to believe that God's ultimate hope for humanity is peace and that positive building-up of community.

The first book and chapter of the Bible displays a life loving and creating God, and then declares that we are made in the same image. Jesus summarized His purpose once in stating: "I have come that they may have life and that they may have it more abundantly." (John 10:10b)

Is it possible that those with an awareness of God's goodness and the ultimate goal of abundant life are particularly (even rightfully) indignant whenever life is undermined and destroyed?

When members of my battalion approached me with hearts filled with pain and vengeance, spewing hatred and hostility, they sometimes assumed they had abandoned their faith and gone to a "place" unreachable by a holy God. I often prescribed imprecatory psalms. That is, I opened to places in the Bible where the language and intentions were as brutally honest and harsh as our thoughts, words, and fantasies:

Let their [the enemies] table become a snare before them, and their well-being a trap. Let their eyes be darkened, so that they do not see, and make their loins shake continually. Pour out Your indignation upon them, and let your wrathful anger take hold of them. Let their dwelling place be desolate; let no one live in their tents. **Psalm 69:22-25**

See also Psalm 59, and even the general comforting "life-cycle" Psalm 139, which reveal transparent and troubling threats and appeals for vengeance. So I was able to remind my troubled companions: "You are not the first 'person of faith' to have deep feelings of resentment and vengeance in response to brutality. These are normal reactions to unusual circumstances."

Reactions against death and destruction we had witnessed, reactions even welling up in us because the good God has created us to recognize good and evil. And yet, reactions which—if acted upon or sustained over time—will sow poison and peril on us, our close relationships, our long-term ability to be people of the life and hope, living into that image of God.

A saying reportedly from the 16th century asserts, "It is not a sin to let a bird fly over...but you don't have to let it nest in your hair." It is one thing to react with a real desire for reprisal. It may even be useful to follow the lead of the Psalmist and put pen to paper, giving voice to such threats turning them over to the One who reserves the right to revenge.

Read Hebrews 10:30: "But great harm can come from unchecked fixations on the settling of scores." When the "bird" of prejudice and hatred is indeed allowed to build a nest, lay and hatch eggs of destruction, despair, even death, will not those fledglings eventually fly away and nest with others, perpetrating the bile of bitterness and bigotry?

Retained bitterness, hostility, anger, prejudices, and hatred are hatched, and spring to life in the most troubling ways upon the most undeserving recipients. One thing that caught me off guard in Iraq was the way the members of my unit, women and men "on the same team," took to attacking one another.

Thankfully it wasn't manifest in physical abuse or attack, but it became obvious that our actions and words, most extreme when friends were killed, turned back on one another and dealt further destruction. Another form is the "friendly fire" of abusive behavior upon returning home directed toward family, friends and even ourselves.

So what can be done? Where is the hope of healing and new life? First, remember that you remain firmly in the hand of the One who created and now wills to restore you. The end of Romans 8 declares that nothing can separate us from God. Not "Angels, nor demons, nor powers...Not distress...persecution... danger...nor sword." Not our feelings or harmful intentions born in violence. No, nothing "Shall be able to separate us from the love of God which is in Christ Jesus our Lord."

Next, remember that this life-loving God had prepared and deployed people of hope and healing: listen to a loved one. Seek out a Chaplain, counselor, doctor, priest, pastor, or other professional. Hatred, prejudice, revenge, whether experienced in Anbar province, or Alaska, or Kirkuk (Iraq) or Keokuk (Iowa) sap strength and vitality. True, God can and will hold us even if we remain in the land of death and darkness, but we will not thrive and experience the intended fullness and abundance of life. We cannot prevent the reactions of the "bird flying overhead." But we can, by the grace of God and the power of Christ, and the healing counsel and community, prevent or eradicate the "nest" and experience transformations and renewed life.

DISCUSSION QUESTIONS

THE FOLLOWING DISCUSSION QUESTIONS ARE BASED ON THE SESSION OVERVIEW, THE ARTICLE, "HATRED, PREJUDICE, AND REVENGE," AND KEY SCRIPTURES FOR THIS SESSION.

1. To what extent do you think the expression of hatred, prejudice, and revenge is a normal reaction to extreme circumstances such as the deaths of teammates in combat? Have you ever acted like this or observed it in others? How was it addressed or what were the results?

2. The author suggests the use of imprecatory psalms in managing one's emotions of revenge or feelings of hatred (Psalm 69:22-25). An imprecatory prayer such as we have in Psalm 69, is a plea for God to pour out His wrath and punishment upon the guilty. How does this square with your understanding of vengeance or the proper expression of anger? Do we find any additional insight from the Key Scriptures?

3. According to the article, how do feelings of hatred and revenge negatively affect the warrior? What are some ways to manage those feelings positively?

4. God is opposed to acts of revenge and expressions of prejudice or hatred. According to Key Scriptures, what does God say about these things, and what does God expect from the obedient warrior who must fulfill his duty, duty to God and duty to complete the mission?

5. What are some dangers of fixating on settling scores, or as the writer says, letting the "bird of prejudice and hatred nest in your hair"? How does God want us to deal with these dangers?

6. The writer refers to the "hope of healing and new life." What does he say about that? How is this helpful in fulfilling one's duty in avoiding expressions of revenge, hatred, or prejudice?

AFTER ACTION REVIEW

1. What new information did you learn about fulfilling your duty concerning the expression of hatred, prejudice, and revenge?

2. What are the dangers of giving in to feelings of hatred, revenge, or prejudice? What does God say about these things?

3. What steps can a warrior take to avoid the unrestrained expressions of hatred and revenge?

4. How has this session helped you in your spiritual journey?

PERSONAL REFLECTION

1. Have you ever felt feelings of hatred, revenge, or prejudice while serving in combat or elsewhere? How did you deal with those feelings? Are you still struggling with some of those feelings?

2. Review the Key Scriptures and identify at least two steps you can take now or in the future to reduce the expression of hatred, revenge, or prejudice.

3. How has this session helped you grow in your faith or your understanding of what God expects of you?

LET'S GO DEEPER

1. Review the Key Scriptures. Count the number of times the word "evil" occurs.
2. How many occurrences did you find?
3. What do the repeated occurrences of this word suggest to you about the topic of hatred, prejudice, and revenge?

REVIEW ROMANS 15:1-3.

1. According to this passage, what is our duty?
2. What does this duty mean to you?

REVIEW LEVITICUS 19:15-18.

1. Identify the "you shall not" commands given in the passage:

   You shall do no ____________________.

   You shall not be ___________ to the poor or _____________ to the great.

   You shall not go around as a ____________________________.

   You shall not _______________ up against the life of your neighbor.

   You shall not _______________ your brother.

   You shall not take ______________________or bear a ______________.

NOTES

**DO YOUR DUTY IN ALL THINGS.
YOU CANNOT DO MORE.
YOU SHOULD NEVER WISH TO DO LESS.**

ROBERT E. LEE

SESSION SEVEN

Resilience

SESSION OVERVIEW:

Resilience is a learned behavior involving both thoughts and actions and includes a process of rapid adaptation in the face of dire circumstances, trauma, general adversity, and other sources of stress. Adaptation is evident when the person sufficiently recovers and returns to her or his original level of functioning.

Resilience is also the ability of an object to regain its shape after being bent, stretched, or compressed. It's always evident in people who are flexible, capable of adjusting to their environment and returning to normal functioning. In this session, we'll take a look at resilience as it mainly relates to persistence. You'll read about Cyrus Field who would not quit despite repeated failures and setbacks. Life isn't any different. Our journey is full of setbacks, failures, and disappointments. But resilient people don't quit. They're people who learn from their failures and rejoice in their challenges. They're also people of faith, who've learned to trust God in any and every circumstance.

GROWTH OBJECTIVES

- TO GAIN A FULL APPRECIATION FOR THE MEANING AND APPLICATION OF RESILIENCE.
- TO EXPLORE WAYS TO ADAPT AND SUCCEED REGARDLESS OF CIRCUMSTANCES.
- TO FIND ENCOURAGEMENT AND INSPIRATION FROM THOSE WHO'VE REFUSED TO QUIT.

KEY SCRIPTURES

1 CORINTHIANS 16:13; EPHESIANS 6:10

Be watchful, stand firm in the faith, act like men, be strong.

Finally, be strong in the Lord and in the strength of his might.

1 CORINTHIANS 10:12-13

Therefore let anyone who thinks that he stands take heed lest he fall. No temptation has overtaken you that is not common to man. God is faithful, and he will not let you be tempted beyond your ability, but with the temptation he will also provide the way of escape, that you may be able to endure it.

2 CORINTHIANS 4:7-10, 16-18a

But we have this treasure in jars of clay, to show that the surpassing power belongs to God and not to us. We are afflicted in every way, but not crushed; perplexed, but not driven to despair, persecuted, but not forsaken; struck down, but not destroyed; always carrying in the body the death of Jesus, so that the life of Jesus may also be manifested in our bodies.

So we do not lose heart. Though our outer self is wasting away, our inner self is being renewed day by day. For this light momentary affliction is preparing for us an eternal weight of glory beyond all comparison, as we look not to the things that are seen but to the things that are unseen.

PHILIPPIANS 3:13-14

Brothers, I do not consider that I have made it my own. But one thing I do: forgetting what lies behind and straining forward to what lies ahead, I press on toward the goal for the prize of the upward call of God in Christ Jesus.

ROMANS 12:2

Do not be conformed to this world, but be transformed by the renewal of your mind, that by testing you may discern what is the will of God, what is good and acceptable and perfect.

PHILIPPIANS 4:11-13

Not that I am speaking of being in need, for I have learned in whatever situation I am to be content. I know how to be brought low, and I know how to abound. In any and every circumstance, I have learned the secret of facing plenty and hunger, abundance and need. I can do all things through him who strengthens me.

KEY SCRIPTURES CONT'D:

2 TIMOTHY 1:7; 2:3

For God gave us a spirit not of fear but of power and love and self-control.

Share in suffering as a good soldier of Christ Jesus.

JAMES 1:12

Blessed is the man who remains steadfast under trial, for when he has stood the test he will receive the crown of life, which God has promised to those who love him.

ROMANS 5:1-5

THEREFORE, SINCE WE HAVE BEEN JUSTIFIED BY FAITH, WE HAVE PEACE WITH GOD THROUGH OUR LORD JESUS CHRIST. THROUGH HIM WE HAVE ALSO OBTAINED ACCESS BY FAITH INTO THIS GRACE IN WHICH WE STAND, AND WE REJOICE IN HOPE OF THE GLORY OF GOD. MORE THAN THAT, WE REJOICE IN OUR SUFFERINGS, KNOWING THAT SUFFERING PRODUCES ENDURANCE, AND ENDURANCE PRODUCES CHARACTER, AND CHARACTER PRODUCES HOPE, AND HOPE DOES NOT PUT US TO SHAME, BECAUSE GOD'S LOVE HAS BEEN POURED INTO OUR HEARTS THROUGH THE HOLY SPIRIT WHO HAS BEEN GIVEN TO US.

WHEN YOU FAIL AGAIN AND AGAIN

THE WARRIOR'S JOURNEY TEAM

THE LORD IS MY HELPER; I WILL NOT FEAR. WHAT CAN MAN DO TO ME? HEBREWS 13:5-6

How many failures should a person allow himself in pursuit of success with a project? How many times should someone fail before he or she calls it quits? Just how many times can success elude him or her before we consider that person a failure? Consider the story of Cyrus W. Field and his efforts to lay the first Trans-Atlantic Cable.

THE FIRST TRY

Cyrus Field cooked up the idea to lay a telegraph cable across the Atlantic Ocean from Great Britain to the United States as early as 1854. By March 1857, he had convinced both governments to invest millions into the project, and by July of that year, work was ready to begin. Both countries agreed to lend large frigates for the task of laying the 2,300-mile long, 2,500-ton cable. The U.S.'s Niagara and the UK's Agamemnon began to lay the cable when disaster struck on this first attempt. Three hundred thirty-five miles away from the coast of Ireland, the Niagara's section of the cable broke, and $500,000-worth of cable plummeted into the ocean. It was a staggering sum of money for that time.

THE SECOND TRY

A year later both ships tried again, using better equipment and a different method. This time they met midway in the Atlantic spliced their cables and headed in opposite directions, the Agamemnon toward the United Kingdom and the Niagara toward the United States. After repeated breaks and the loss of hundreds of miles of cable, the two ships returned in defeat to Great Britain and a furious public outcry.

In despair, numerous members of the project resigned. At a gloomy meeting of the remaining partners, Fields and his associates reluctantly agreed to try one more time to lay the 2,300-mile-long cable. Initially, it seemed like they had finally succeeded. By August of 1858, Cyrus Field completed the long stretch from Ireland to Newfoundland and tapped out the world's first trans-Atlantic telegraph message.

But Cyrus's success did not last long. At the worst possible time, the cable suddenly ceased to function. It was right in the middle of a banquet in Cyrus Field's honor. The United States, the United Kingdom, and their public were outraged and demanded a thorough investigation. By this time the project resulted in wasted millions of dollars and had nothing to show for it.

THE THIRD TRY

Another seven years and a civil war passed before Cyrus revived the project again. This time Cyrus employed a much improved, twice as thick and twice as heavy, cable and the large steamer, the Great Eastern. Cyrus Field made this latest attempt in July 1865. Though exceedingly optimistic, Cyrus Field saw this effort end in disaster as well.

THE FOURTH TRY

What did Cyrus do? Call it quits? Resign himself to failure? No, he took all the lessons he learned from every failure and used the knowledge to improve on his methods and his equipment. One year later, on July 27, 1866, Cyrus Field tapped out this message across the Atlantic: "All well. Thank God, the cable is laid and in perfect working order!"

Now, answer the question. Does history regard Cyrus W. Field as a flop, as a failure? Not at all. History only records his success. But if he had quit while he was behind, how would history have judged him? You see, persistence and the refusal to accept defeat make the difference between dismal failure and stunning achievement. God promises us in his word, "I will never leave you nor forsake you." So we may boldly say: 'The Lord is my helper; I will not fear. What can man do to me?' (Hebrews 13:5-6, NKJV)

DISCUSSION QUESTIONS

THE FOLLOWING DISCUSSION QUESTIONS ARE BASED ON THE SESSION OVERVIEW, THE ARTICLE, "WHEN YOU FAIL AGAIN AND AGAIN," AND KEY SCRIPTURES FOR THIS SESSION.

1. How do you usually handle failure or disappointment? How do you respond to adversity?

2. God often uses failures or trials to test us. Can you think of any recent experiences where you may have been tested? How did you do? What did you learn?

3. Paul refused to quit despite repeated setbacks and outright assaults. Why do you think that was the case? What kept Paul focused? How do you think Paul's sense of purpose propelled him forward? How is Paul's situation similar to Cyrus Field's?

4. Resilience is the ability or capacity to bounce back quickly from adversity and regain normal functioning. What factors or attitudes help you do this? What kinds of things inhibit resilience?

AFTER ACTION REVIEW

1. What new information did you learn about developing resilience?

2. Identify the factors that contribute to resilience in a person?

3. How has this session helped you in your spiritual journey?

PERSONAL REFLECTION

1. What are the things that undermine your resilience?

2. How has failure affected your ability to persevere? Can you think of situations or circumstances that led to failure but you eventually overcame and found success? What helped you get beyond your failures? What did you learn about yourself through the process?

3. How has this session contributed to your spiritual growth?

LET'S GO DEEPER

REVIEW 2 CORINTHIANS 4:7-10, 16-18a.

1. What do you think was the secret of Paul's resilience according to the passage?

2. How can you apply this "secret" to your circumstances?

REVIEW PHILIPPIANS 4:11-13.

1. In what ways does the apostle reveal his spiritual flexibility?

   He learned in whatever situation to be ________________.

   He knew how to be brought __________ and how to ______________.

   In any and every circumstance, he learned the ____________ of facing plenty and hunger, abundance and need.

2. What do you think was the key to that secret?

NOTES

NO ONE ESCAPES PAIN, FEAR, AND SUFFERING. YET FROM PAIN CAN COME WISDOM, FROM FEAR CAN COME COURAGE, FROM SUFFERING CAN COME STRENGTH -- IF WE HAVE THE VIRTUE OF RESILIENCE.

ERIC GREITENS

SESSION EIGHT

Leadership

SESSION OVERVIEW:

Leadership is defined as influencing the thinking, behavior, and development of another person. Jeremy Stalnecker, in *Leadership By Design*, believes it's merely "taking someone from where they are to where they need to be." In this session, we'll consider the attributes of an effective leader, and what it means to lead. Leadership starts with the individual and grows out of a person's character. The Key Scriptures reveal that good leaders are men and women who fear God, depend on God, and serve others. Leaders model exemplary behavior. They lead by example, provide direction, motivate others to excel and, when necessary, come alongside them to provide the support they need. Leaders are never about themselves. True leaders always think about others first.

GROWTH OBJECTIVES

- TO IDENTIFY THE KEY ELEMENTS OF EFFECTIVE LEADERSHIP.
- TO ACQUIRE AN UNDERSTANDING OF WHAT IT MEANS TO BE A SERVANT-LEADER.
- TO ADOPT ATTITUDES AND BEHAVIORS TO IMPROVE LEADERSHIP CAPABILITY.

KEY SCRIPTURES

MATTHEW 20:25-28

But Jesus called them to him and said, 'You know that the rulers of the Gentiles lord it over them, and their great ones exercise authority over them. It shall not be so among you. But whoever would be great among you must be your servant, and whoever would be first among you must be your slave, even as the Son of Man came not to be served but to serve, and to give his life as a ransom for many.'

EXODUS 18:21

Moreover, look for able men from all the people, men who fear God, who are trustworthy and hate a bribe, and place such men over the people as chiefs of thousands, of hundreds, of fifties, and of tens.

DEUTERONOMY 1:13

Choose for your tribes wise, understanding, and experienced men, and I will appoint them as your heads.

2 CHRONICLES 1:10-12a

Give me now wisdom and knowledge to go out and come in before this people, for who can govern this people of yours, which is so great?" God answered Solomon, 'Because this was in your heart, and you have not asked possessions...but have asked wisdom and knowledge for yourself that you may govern...wisdom and knowledge are granted to you.

JOHN 13:12-15

When he had washed their feet and put on his outer garments and resumed his place, he said to them, 'Do you understand what I have done to you? You call me Teacher and Lord, and you are right, for so I am. If I then... have washed your feet, you also ought to wash one another's feet. For I have given you an example, that you also should do just as I have done to you.'

HEBREWS 13:7, 17

Remember your leaders...Consider the outcome of their way of life, and imitate their faith.

Obey your leaders and submit to them, for they are keeping watch over your souls, as those who will have to give an account. Let them do this with joy and not with groaning, for that would be of no advantage to you.

1 TIMOTHY 3:1-7

The saying is trustworthy: If anyone aspires to the office of overseer, he desires a noble task. Therefore an overseer must be above reproach, the husband of one wife, sober-minded, self-controlled, respectable, hospitable, able to teach, not a drunkard, not violent but gentle, not quarrelsome, not a lover of money. He must manage his own household well, with all dignity keeping his children submissive, for if someone does not knowhow to manage his own household, how will he care for God's church?

PHILIPPIANS 2:3-4

DO NOTHING FROM RIVALRY OR CONCEIT, BUT IN HUMILITY COUNT OTHERS MORE SIGNIFICANT THAN YOURSELVES. LET EACH OF YOU LOOK NOT ONLY TO HIS OWN INTERESTS, BUT ALSO TO THE INTERESTS OF OTHERS.

THE BUCK STOPS HERE

CHAPLAIN (COLONEL) SCOTT MCCHRYSTAL, U.S. ARMY (RET)

HE DEMONSTRATED COURAGE AND ACCOUNTABILITY FOR HIS ACTIONS...HE SHOWED MORE CONCERN FOR OTHERS THAN FOR HIMSELF.

LEADERSHIP IDENTITY

Realistic training for combat can have many moving parts and tends to get complicated fast. This is especially true when using live ammunition. Leaders know they must fashion good plans, issue clear orders, and delegate authority for missions to succeed. Ultimately, however, the responsibility falls on the shoulders of the senior leader.

Mixed in with this is the issue of safety. Training needs to tax the troops to the limit, but there is a line that shouldn't be crossed to avoid unnecessary casualties. Risk assessment is not easy.

In spite of all precautions, stuff happens. Things go wrong, and people get hurt or killed. Times like this reveal a lot about the leaders. Competent and caring leaders step in and take charge. They want to take all the necessary steps to rectify the situation and take care of troops.

Self-serving leaders tend to go into a defensive mode to protect themselves. Blaming others is a common tactic. They want to protect their careers at all costs.

In the early 1970s, I belonged to a unit that deployed to a Caribbean island for a valuable training exercise. I was placed in charge of the rear deployment while the unit deployed for approximately two weeks.

FRIENDLY FIRE

Things seemed to be running smoothly. Daily I would receive reports about the training. Late one evening, however, I received a disturbing report that a severe training accident had occurred in which several soldiers had been killed or wounded.

During a live-fire exercise in which mortar rounds were being fired, several rounds hit in the vicinity of soldiers as they moved on foot to the objective. Evidently, a mistake resulted in bullets being shot onto the wrong location. It was a tragic incident in which "friendly fire" proved fatal. The unit took all the necessary steps to evacuate the wounded quickly and get them to proper medical care. Unfortunately, several soldiers didn't make it.

Back at Fort Bragg, we initiated action to notify family members of soldiers killed or wounded. We took extra precautions to disseminate information quickly, but accurately. Wrong information in cases like this only compounds the problems.

The division headquarters sent safety experts to join the unit to conduct a thorough investigation of the circumstances surrounding the training accident. They interviewed dozens of people, gathered hundreds of pages of depositions, and took all measures necessary toward determining the causes of the deadly incident. The unit returned, but the investigation continued for weeks after the troops had arrived back at home station.

THE BACKGROUND

From the first day that the unit returned, I heard a lot of talk about the accident from leaders and soldiers at all levels. Understandably, there was much speculation as to where the mistakes had occurred. No one person had all of the information, and everyone anxiously awaited the report from the investigation. The report would likely find the causes for the accident and affix responsibility for the mishap. Most likely, those assigned the blame would face serious consequences.

Human nature kicked in quickly as the speculation continued. I heard some leaders talking in ways to protect themselves and their careers. They repeatedly rehearsed the measures they had taken to ensure that the exercise had gone safely. I didn't hear anyone even hint that they might have contributed to the accident.

COURAGE

After several more weeks, word traveled throughout the unit that the report was final and rested in the hands of the Division Commander for the last review. But then an interesting thing happened. Before top leaders released any specifics from the investigation, our battalion commander stepped forward publicly and accepted responsibility for the incident. He stated that since he was the senior leader on the ground, he rightfully was responsible for everything that happened or failed to happen. In effect, he said: *The buck stops here.*

The hammer didn't take long to fall. Our battalion commander was relieved for the incident and moved to another duty assignment. To my knowledge, no other officer or non-commissioned officer received punishment for the accident. The senior leader took the entire brunt of the punishment on his shoulders. It seemed unfair. Yes, he deserved a measure of punishment, but not all of it!

SERVANT LEADERSHIP

I remember my amazement that this man would step up and take the entire blame without trying to fault others. Unlike some less senior leaders in the unit, he demonstrated courage and accountability for his actions. Perhaps most inspiring of all, he showed more concern for others than for himself. In spite of the negative impact on his own life and career, he chose to sacrifice himself that others would not have to pay.

Decades have passed, and I still think about this leader often. He has my utmost respect and admiration. I can also tell you this about him. I would serve under him again—any time, any place. He's a leader I can follow.

Does this sound like any other leader you know? Almost 2,000 years ago, Jesus Christ descended from heaven as a man. He surrendered many of the privileges of divinity to become like us. He came and showed us how to live. More amazingly, he came to die for our sins so that the Father's judgment for our sins would not fall on us. To put it more directly, Jesus died for your sins and mine. And He has asked us to follow Him. Will you?

DISCUSSION QUESTIONS

THE FOLLOWING DISCUSSION QUESTIONS ARE BASED ON THE SESSION OVERVIEW, THE ARTICLE, "THE BUCK STOPS HERE," AND KEY SCRIPTURES FOR THIS SESSION.

1. Given the Session Overview, Key Scriptures, and article, what are the critical elements of leadership? How would you describe a good leader?

2. The Services often talk about "mission first, people always." What does that mean to you and how does that reflect effective leadership?

3. How is the decision by the battalion commander to accept full responsibility for the training deaths commendable? What were the consequences of his choice? Do you agree or disagree with how it was handled? Why?

4. How would you describe "servant leadership"? Do you think servant leadership, commended in Scripture and the article, is the appropriate model for military leadership? How so? Can you think of any examples of servant leaders? Were they successful? Why?

AFTER ACTION REVIEW

1. What new information did you learn about leading others?

2. What does it mean to be a servant-leader?

3. How has this session helped you in your spiritual journey?

PERSONAL REFLECTION

1. What is your leadership style and how might you adjust it given the information you've learned in this session?

2. Good leaders are ready to accept blame when things go wrong and give credit to others when things go well. What is your first inclination when things go wrong?

3. Jesus washed the feet of His followers. What was He trying to show them about leading others?

LET'S GO DEEPER.

REVIEW MATTHEW 20:25-28.

1. In this passage, Jesus contrasts the leadership style of the Gentiles with what He expects of His followers. How is Jesus' idea of leadership different?

2. What do you think Jesus meant when he said "*The rulers of the Gentiles lord it over them*"?

REVIEW EXODUS 18:21 & DEUTERONOMY 1:13.

1. Identify the attributes God expects of competent leaders.

   People who are _________

   People who ________ God

   People who are ___________________________

   People who hate a ___________________

   People who are ____________, _____________, and _____________

LET'S GO DEEPER CONT'D:

REVIEW 1 TIMOTHY 3:1-7.

1. Underline all the attributes of a church leader in the passage:

   The saying is trustworthy: If anyone aspires to the office of overseer, he desires a noble task. Therefore an overseer must be above reproach, the husband of one wife, sober-minded, self-controlled, respectable, hospitable, able to teach, not a drunkard, not violent but gentle, not quarrelsome, not a lover of money. He must manage his own household well, with all dignity keeping his children submissive, for if someone does not know how to manage his own household, how will he care for God's church?

2. What are the implications for a warrior-leader?

3. Why do you think God focuses so much on character qualities instead of competencies?

NOTES

LEADERSHIP IS NOT ABOUT TITLES, OR POSITIONS. IT IS ABOUT ONE LIFE INFLUENCING ANOTHER.

JOHN C. MAXWELL

SESSION NINE

Faithfulness

SESSION OVERVIEW:

"Always Faithful." *Semper Fidelis* or *Semper Fi*. It serves as the motto for the United States Marine Corps. Marines live or die by those words. They speak of loyalty, duty, and sacrifice. The words are sacred because they were born in blood, through the sacrifice of all who have served.

God has also called every believer to faithfulness. He is faithful and expects no less from us. In this session, we'll consider an article titled "Two Men, Two Choices," and Key Scriptures that speak to faithfulness and the choices people make. Decisions have a lot to do with faithfulness. Our choices are like railroad switches. The choices we make determine where we end up. Like Marines, who choose to live or die by their motto, followers of Jesus must choose faithfulness over faithlessness, commitment over fecklessness, and truth over falsity. Indeed, faithfulness must infuse one's entire life, in attitude and every action, word and deed.

GROWTH OBJECTIVES

- TO GRASP THE IMPORTANCE OF FAITHFULNESS IN LIVING THE CHRISTIAN LIFE AND IN SERVICE TO OTHERS.
- TO RECOGNIZE THE RELATIONSHIP FAITHFULNESS HAS TO THE CHOICES WE MAKE.
- TO COMMIT TO FAITHFULNESS IN ATTITUDE AND ACTION, WORD, AND DEED.

KEY SCRIPTURES

PROVERBS 1:7, 29-31

The fear of the LORD is the beginning of knowledge; fools despise wisdom and instruction.

Because they hated knowledge and did not choose the fear of the LORD, would have none of my counsel and despised all my reproof; therefore they shall eat the fruit of their way, and have their fill of their own devices.

JOSHUA 24:14-15

Now, therefore, fear the LORD and serve him in sincerity and in faithfulness. Put away the gods that your fathers served beyond the River and in Egypt, and serve the LORD. And if it is evil in your eyes to serve the LORD, choose this day whom you will serve...But as for me and my house, we will serve the LORD.

DEUTERONOMY 7:9

Know therefore that the LORD your God is God, the faithful God who keeps covenant and steadfast love with those who love him and keep his commandments, to a thousand generations.

MATTHEW 16:24

Then Jesus told his disciples, 'If anyone would come after me, let him deny himself and take up his cross and follow me.'

HEBREWS 3:1b-2, 5-6a

Consider Jesus...who was faithful to him who appointed him, just as Moses also was faithful in all God's house.

Now Moses was faithful in God's house as a servant, to testify to the things that were to be spoken later, but Christ is faithful over God's house as a son.

HEBREWS 10:23, 11:6

Let us hold fast the confession of our hope without wavering, for he who promised is faithful.

And without faith, it is impossible to please him, for whoever would draw near to God must believe that he exists and that he rewards those who seek him.

LAMENTATIONS 3:21-23

BUT THIS I CALL TO MIND, AND THEREFORE I HAVE HOPE: THE STEADFAST LOVE OF THE LORD NEVER CEASES; HIS MERCIES NEVER COME TO AN END; THEY ARE NEW EVERY MORNING; GREAT IS YOUR FAITHFULNESS.

TWO MEN, TWO CHOICES

CHAPLAIN (COLONEL) SCOTT MCCHRYSTAL, U.S. ARMY (RET)

THEN PETER REMEMBERED THE WORD JESUS HAD SPOKEN: 'BEFORE THE ROOSTER CROWS, YOU WILL DISOWN ME THREE TIMES.' AND HE WENT OUTSIDE AND WEPT BITTERLY. *MATTHEW 26:75*

CHOICES

"When Judas, who had betrayed him, saw that Jesus was condemned, he was seized with remorse and returned the thirty silver coins to the chief priests and the elders. 'I have sinned,' he said, 'for I have betrayed innocent blood.' 'What is that to us?' they replied. 'That's your responsibility.' So Judas threw the money into the temple and left. Then he went away and hanged himself" (Matthew 27:3-5). Life offers us choices, and choices matter because they impact others and ourselves.

One relevant category of choices relate to the decisions we make during times of stress, pressure, and crisis. Typically, people are not at their strongest in these situations. They are not thinking, don't have their rational perspective, and they often let their emotions override sound judgment. The results of choices range from inconsequential to major, from immediate to long-range, from temporary to eternal. But the bottom line is simple: good decisions are better than poor ones.

CONTEXT

I recall a young lady at one military installation who came to see me. She had entered into a relationship with a young officer on post and made a poor decision that resulted in her becoming pregnant. But that was hardly the end of her stress. Although marriage was out of the question, the officer voluntarily offered child support once the baby was born.

However, he changed his mind. He was not going to help her financially and now encouraged her to terminate her pregnancy. The young lady's world had been turned upside down, but she didn't lose her head. I encouraged her not to terminate her pregnancy and helped her explore options once the baby was born. She prayed to the Lord for wisdom and guidance. She chose to have the baby.

In the months following the birth of her daughter, she came to see me a couple of times. She thanked me many times for my assistance, particularly my encouragement to her to seek God's guidance in the whole matter. She beamed with joy and happiness. Though she had made one poor decision that created significant stress in her life, she responded in a godly manner. She made the right choice. Making sound decisions during challenging times is not easy, nor is it automatic.

WISDOM

The ability to decide wisely doesn't necessarily correlate to a person's age, experience, education, or financial status. All of these can play significantly into one's ability to make choices, but they are not determinate. We always need God's help, but particularly during times in our lives when stressful conditions seem overwhelming to us. We need to know that God cares, that He understands and that He loves us unconditionally.

We also want assurance that God has the clout to help us—that nothing is too complicated for Him. Sometimes, our circumstances may appear so dark that we don't even know how we will keep going. Hopelessness, depression, lack of purpose—God can deal with any of these issues, and much more. Perhaps above anything else the Lord provides for us, we need His wisdom. The good news is that God's wisdom is available.

The Bible tells us this in James 1:5: "If any of you lacks wisdom, he should ask God, who gives generously to all without finding fault, and it will be given to him." The bad news, however, is that in desperate circumstances many people don't seek it. They go after information from other sources and often wind up making the wrong choice. There is one vital reality regarding the choices we make, and many still haven't accepted the truth on this one. Regardless of the situation and how hard it may be, the individual has to assume responsibility for the choice.

RESPONSIBILITY

Yes, there can be enormous pressure from many sides, but ultimately we make the decision, not someone else. Why is this so crucial? This is why. If we continually blame the circumstances, another person, or cite something else as the reason for the choice, we never learn to take responsibility for our actions. We don't learn to seek the truth and deal squarely with the facts. We never develop our God-given potential as a human being to behave in ways that honor God.

In the final analysis, God will judge each of us for the things we have done or left undone. It makes a lot more sense to understand this now and start accepting accountability for our choices. God is just and fair. But the Bible makes it clear that He will hold each of us accountable for our decisions.

By this point, I trust you are more persuaded to seek the Lord during your trials. Even if you are the culprit and brought the circumstances upon yourself, our gracious God is available to help. But the choice to seek his support belongs to you.

Let's conclude this discussion about choices by contrasting two biblical figures. Each encountered an incredibly robust set of circumstances. Each had to make a choice. As we examine their situations along with the decisions, each made, note the stark contrast in the consequences of their choices.

REAPING AND SOWING

The two men are Peter and Judas, both among Jesus' original twelve disciples. Both men were part of the select group that traveled with Jesus during his three years of ministry. Both were recipients of his teaching and witnessed the many miracles Jesus did. Both were present with Jesus in the upper room when the Master shared one final meal with his followers. During the time of Jesus' arrest and mock trial, both found themselves in the most stressful circumstances of their lives.

Consider Judas, the disciple whose betrayal prompted the religious leaders to arrest Jesus. After a long night of inquisition, Jesus received the death sentence. Scripture tells us that at this point Judas realized his wrongdoing and went back to the religious leaders to confess his sin. Despite Judas' admission, the decision to kill Jesus still stood. It's impossible to put ourselves in the shoes of the very man who betrayed Jesus Christ. But we can examine the breadth of Scripture and conclude that Judas' decision to take his own life was not his only choice. His was a fateful decision with eternal consequences. Nothing in the Bible provides evidence that Judas made it to heaven. The evidence seems to indicate just the opposite—that Judas would spend eternity separated from God.

PRESSURE

It was the wrong choice under unimaginably stressful conditions. Could

Judas have sought God for forgiveness? Could he have reset the course of his life and continued serving as a disciple of Christ? The answer is a resounding 'yes.' But that's not the choice Judas made. It was his decision, and he bears full responsibility for it.

Let's review Peter's situation. Leading up his arrest, Jesus had told Peter that the disciple would deny Him three times. Unquestionably, Peter didn't believe it. But following Jesus' arrest, things started to unravel for the brash disciple. On three separate occasions, Peter denied ever having been with Jesus. Following his last denial, the rooster crowed just as Jesus had foretold. Peter was both stunned and shocked by his lack of courage.

All of his claims about remaining loyal to Jesus, even to his death, were nothing but hot air. During the most crucial moments to date, Peter had failed his Lord. What could have been worse? Like Judas in many respects, Peter had failed Jesus.

FORGIVENESS

But unlike Judas, Peter exercised a radically different choice. Instead of deciding to take his own life, he doubtlessly sought forgiveness and direction from God through prayer. Only Peter and God alone could know how painful the next few days must have been. Jesus, their leader, gets crucified and buried.

Peter, along with the other followers, had no idea of what was coming next. News of the Resurrection must have amazed Peter, but not nearly as much as the appearances, Jesus made to the disciples following his rising from the dead. The most special occurred when Jesus appeared to Peter and a few of the disciples along the shore of the Sea of Galilee.

At that time, the Gospel of John records pieces of a conversation between Jesus and Peter. The essence of the discussion focused around Jesus telling Peter to continue his ministry. Phrases like 'feed my sheep' and 'follow me' gave new meaning and purpose to Peter's life. The rest is history.

Peter went on to become the recognized leader of the disciples. He even wrote two books of the Bible—First and Second Peter. Until his death, Peter's life was committed to preaching the Gospel and building the Church. Though not fully substantiated, Christian tradition reports that Peter died a martyr's death, being crucified upside down. The upside down part was at Peter's request, for he didn't feel worthy of being crucified in the same manner as his Lord. Two men. Two choices. Seek God. Make the right choices.

DISCUSSION QUESTIONS

THE FOLLOWING DISCUSSION QUESTIONS ARE BASED ON THE SESSION OVERVIEW, THE ARTICLE, "TWO MEN, TWO CHOICES," AND KEY SCRIPTURES FOR THIS SESSION.

1. The choices we make determine how we live and where we end up in life. Why is this true and how do choices relate to faithfulness?

2. Chaplain McCrystal says the stresses and pressures of life often have a lot to do with the choices we make. Like the young woman who was faced with terminating her pregnancy or keeping the baby, pressures like this can distort perspective and influence decisions. How have stressful events or circumstances affected your choices? What were your options? What did you decide?

3. Two men were faced with similar circumstances. Judas betrayed Jesus probably for the money. Peter denied Jesus because of fear. Both represented a betrayal. But the outcome for these men was different. Why was that? What can you learn from this?

4. God calls us to faithfulness. He is faithful and expects no less from us. What can you do to nurture faithfulness in your daily life? How does this question relate to obedience, commitment, and determination?

AFTER ACTION REVIEW

1. What new information did you learn about faithfulness?

2. Why is taking responsibility for your actions so important? How does this relate to faithfulness and the choices you make?

3. God grants wisdom to us in times of stress. When we fail, he offers forgiveness. How do these things relate to faithfulness and encourage responsible and godly living?

4. How has this session helped you in your spiritual journey?

PERSONAL REFLECTION

1. Reflect on a time when you were unfaithful, less committed, or made poor choices. What factors influenced you the most? How did you recover?

2. When faced with stresses and pressures, how does God want you to respond?

3. How has this session helped you grow in your faith or your understanding of what God expects of you?

LET'S GO DEEPER

REVIEW JOSHUA 24:14-15.

1. Complete the verse by filling in the blanks:

   Now therefore __________ the LORD and serve him in ______________ and in ____________________. Put away the gods that your fathers served beyond the River and in Egypt, and ____________ the LORD. And if it is evil in your eyes to serve the LORD, ______________ this day whom you will serve, . . . But as for me and my house, we will serve the LORD.

2. What does this passage say about Joshua's resolve?

3. What were their choices?

   To serve the gods _____________ or to serve the ______________.

REVIEW DEUTERONOMY 7:9.

1. How does this passage describe God's faithfulness?

REVIEW HEBREWS 10:23; 11:6.

1. What is the one indispensable ingredient for pleasing God?

2. How does the passage describe faithfulness?

3. What does God promise to those who seek Him?

NOTES

THE GLORY OF GOD'S FAITHFULNESS IS THAT NO SIN OF OURS HAS EVER MADE HIM UNFAITHFUL.

CHARLES SPURGEON

SESSION TEN

Integrity

SESSION OVERVIEW:

The American Heritage Dictionary defines integrity as "steadfast adherence to a strict moral or ethical code...the state of being unimpaired; soundness." It's a "quality or condition of being whole or undivided; completeness." The concept speaks of consistency of substance, uniformed adherence, devoid of foreign or corrupt matter. Moral integrity is consistent adherence to moral principles evident in purity of thought, motive, and action, honesty of character; especially with respect to words and actions. A person of integrity is one whose life is consistently upright publicly and privately, and whose heart is morally and ethically whole.

In the Old Testament, the word is derived from a word that means to be complete and conveys the moral sense of innocence and upright simplicity.

GROWTH OBJECTIVES

- TO UNDERSTAND THE IMPORTANCE AND MEANING OF INTEGRITY.
- TO IDENTIFY AREAS OF WEAKNESS IN THE MAINTENANCE OF PERSONAL AND PUBLIC INTEGRITY.
- TO APPLY CONCEPTS THAT SUPPORT THE DEVELOPMENT OF INTEGRITY.

KEY SCRIPTURES

PSALM 25:21

May integrity and uprightness preserve me, for I wait for you.

PROVERBS 11:3

The integrity of the upright guides them, but the crookedness of the treacherous destroys them.

PROVERBS 19:1

Better is a poor person who walks in his integrity than one who is crooked in speech and is a fool.

PROVERBS 20:7

The righteous who walks in his integrity—blessed are his children after him!

GENESIS 20:5

Now Abimelech had not approached her. So he said, 'Lord, will you kill an innocent people? Did he not himself say to me, 'She is my sister'? And she herself said, 'He is my brother.' In the integrity of my heart and the innocence of my hands I have done this.

1 KINGS 9:4

And as for you[Solomon], if you will walk before me, as David your father walked, with integrity of heart and uprightness, doing according to all that I have commanded you, and keeping my statutes and my rules, then I will establish your royal throne over Israel forever, as I promised David your father...

JOB 2:3

And the LORD said to Satan, 'Have you considered my servant Job, that there is none like him on the earth, a blameless and upright man, who fears God and turns away from evil? He still holds fast his integrity, although you incited me against him to destroy him without reason.'

JOB 27:5-6

Far be it from me to say that you are right; till I die I will not put away my integrity from me. I hold fast my righteousness and will not let it go; my heart does not reproach me for any of my days.

PROVERBS 2:6-8, 21

FOR THE LORD GIVES WISDOM; FROM HIS MOUTH COME KNOWLEDGE AND UNDERSTANDING; HE STORES UP SOUND WISDOM FOR THE UPRIGHT; HE IS A SHIELD TO THOSE WHO WALK IN INTEGRITY, GUARDING THE PATHS OF JUSTICE AND WATCHING OVER THE WAY OF HIS SAINTS.

FOR THE UPRIGHT WILL INHABIT THE LAND, AND THOSE WITH INTEGRITY WILL REMAIN IN IT.

A CULTURE OF HONESTY AND INTEGRITY

THE WARRIOR'S JOURNEY TEAM

LET INTEGRITY AND UPRIGHTNESS PRESERVE ME. PSALM 25:21

Do science and morality share any bond? One country is finding out the hard way that the two are inseparable.

China's President, Xi Jinping declared that his goal is turning China into "a global scientific and technology power" by the year 2049. Unfortunately, there has been a shocking increase in the lack of honesty and integrity by some of its academicians and scientists. This situation is undermining President Jinping's goal and destroying China's credibility with other nations.

A recent New York Times article stated that since 2012 China has led the world in falsified peer reviews of scientific papers. In the last year a single scientific periodical, Tumor Biology, retracted 107 of its published articles. This was after their peer reviews were found to be fabricated; and, Chinese scientists authored almost all of these.

One Chinese geneticist, Han Chunyu, of the Hebei University of Science and Technology, won international fame after publishing an article in Nature Biotechnology. In the article, he claimed to have discovered a method to edit the human genome—in such a way to eliminate diseases and help parents choose their child's gender and IQ. In response to his global notoriety, the "local government even offered to build [Han Chunyu] a $32 million gene-editing research center at his university, which he would run." Alas, it did not come to be. He also faked his research, results and peer reviews.

Why all this lack of honesty? Many blame it on the standard of success, which the government imposes. This standard is based upon the number of scientific papers a researcher gets published (in reputable periodicals) and how many times his or her work other authors cite it. This standard is referred to as the Science Citation Index (SCI). Those with the highest SCI would receive fame and funding for further research.

Unfortunately, the selling of phony peer evaluations and completed research papers is a huge business in China. The Times reported that these organizations offer services ranging from faked peer reviews to entire scientific articles already written and ready to submit. One of the companies advertised, "We have helped professors of all backgrounds. Don't worry, we'll keep it a secret." One respected scientist lamented, "We need to work harder to develop a culture of integrity."

Back in the 1990s, the importance of character, honesty, and integrity was primarily dismissed in America. All that mattered was intelligence and job performance. What a person was and did in his time was nobody else's business. Perhaps this abandonment of morals was an effort to accommodate an administration, which seemed devoid of morality and character?

As China can demonstrate, sound science and academic excellence can only take place with the practice of honesty and integrity. Without morality, science degrades into a culture of cheating and mediocrity.

In Psalm 25:21 King David prayed, "Let integrity and uprightness preserve me" (NKJV).

DISCUSSION QUESTIONS

THE FOLLOWING DISCUSSION QUESTIONS ARE BASED ON THE SESSION OVERVIEW, THE ARTICLE, "A CULTURE OF HONESTY AND INTEGRITY," AND KEY SCRIPTURES FOR THIS SESSION.

1. Integrity is closely aligned with uprightness. How would you describe uprightness? Provide some examples that are relevant to your life and work situation?

2. According to the article, China's technological dishonesty has undermined its credibility on the world stage. How does a pattern of deception affect personal and organizational credibility?

3. Some people say character isn't as important as results. How is this mentality evident or not evident in the military? Why? Why not?

4. What can be done to prevent a culture of cheating and mediocrity in the military? How is that important in a military organization?

5. General George Washington said "Honesty is the best policy," in public as well as private affairs. What do you think this means in today's culture? How does it apply?

6. Integrity is consistent adherence to moral and ethical standards. Given the shifting nature of moral values in society, how does a warrior adhere to what is right in word and action? What challenges present themselves?

AFTER ACTION REVIEW

1. What new information did you learn about integrity?

2. In what ways does the practice of dishonesty undermine organizational credibility? How does it impact the mission?

3. How does a warrior protect himself from the dangers of dishonesty?

4. How has this session helped you in your spiritual journey?

PERSONAL REFLECTION

1. Why is integrity fundamental to living a good life? How do you ensure you are true to self and honest in your dealings?

2. What areas in your professional and private life need some "shoring up"? Where are you vulnerable to dishonesty or mediocrity?

3. How has this session helped you grow in your faith or your understanding of what God expects of you?

LET'S GO DEEPER.

REVIEW 1 KINGS 9:4.

1. How does God describe integrity?

   To _______ before me as David walked.

   Doing according to all that I have _________________ you.

   ____________ my statues and my ____________.

2. What did God promise to Solomon if he walked with integrity of heart?

REVIEW JOB 2:3.

1. How does God describe Job?

   Job is a __________________ and ______________ man.

   Job ________ God.

   Job turns away from ___________.

2. How do these things contribute to a person's integrity?

LET'S GO DEEPER CONT'D:

REVIEW PROVERBS 2:6-8, 21.

1. Complete the following verse by filling in the blanks:

   For the LORD gives ____________; from his mouth come ____________ and ________________; he stores up sound wisdom for the ____________; he is a shield to those who walk in ______________, guarding the paths of justice and watching over the way of his____________.

   For the upright will inhabit the land, and those with ________________ will remain in it.

2. What three things does God give to the upright?

   He gives ________________.

   He gives ________________.

   He gives ________________.

3. What does God promise to do for those who walk in integrity?

NOTES

INTEGRITY IS DOING THE RIGHT THING, EVEN WHEN NO ONE IS WATCHING.

C.S. LEWIS

SESSION ELEVEN

Selfless Service

SESSION OVERVIEW:

In an address to the graduating class of the U.S. Naval Academy in 1961, President John F. Kennedy said: "What you have chosen to do for your country by devoting your life to the service of your county is the greatest contribution that any man could make." Service, selfless service, which is the only kind that matters, is a sacred thing. Whether we speak of serving the nation, one another, or God, selfless service is servanthood. Service is sacrificing oneself for the benefit of others. Service always costs the server. In the final analysis, selfless service is really about loving God and loving people. It is also the path to greatness, as Jesus showed us, laying aside the robes of royalty to take upon Himself the garments of a slave so that He might give Himself completely for you and me.

GROWTH OBJECTIVES

- TO DISCOVER THE SIGNIFICANT FACTORS THAT CONTRIBUTE TO SELFLESS SERVICE.
- TO IDENTIFY THE THINGS THAT NEGATIVELY AFFECT SELFLESS SERVICE.
- TO RENEW ONESELF TO THE TASK OF SERVING SELFLESSLY.

KEY SCRIPTURES

MATTHEW 18:1-4

At that time the disciples came to Jesus, saying, 'Who is the greatest in the kingdom of heaven?' And calling to him a child, he put him in the midst of them and said, 'Truly, I say to you, unless you turn and become like children, you will never enter the kingdom of heaven. Whoever humbles himself like this child is the greatest in the kingdom of heaven.

MATTHEW 20:26-28

It shall not be so among you. But whoever would be great among you must be your servant, and whoever would be first among you must be your slave, even as the Son of Man came not to be served but to serve, and to give his life as a ransom for many.

MATTHEW 22:37-38

And he said to him, 'You shall love the Lord your God with all your heart and with all your soul and with all you mind. This is the great and first commandment. And a second is like it: You shall love your neighbor as yourself.

LUKE 22:26-27

But not so with you. Rather, let the greatest among you become as the youngest, and the leader as one who serves. For who is the greater, one who reclines at table or one who serves? Is it not the one who reclines at table? But I am among you as the one who serves.

LUKE 10:33-34

But a Samaritan, as he journeyed, came to where he was, and when he saw him, he had compassion. He went to him and bound up his wounds, pouring on oil and wine. Then he set him on his own animal and brought him to an inn and took care of him.

JOHN 13:5, 14

Then [Jesus] poured water into a basin and began to wash the disciples' feet and to wipe them with the towel that was wrapped around him.

'If I then, your Lord and Teacher, have washed your feet, you also ought to wash one another's feet.'

PHILIPPIANS 2:3-7a

DO NOTHING FROM RIVALRY OR CONCEIT, BUT IN HUMILITY COUNT OTHERS MORE SIGNIFICANT THAN YOURSELVES. LET EACH OF YOU LOOK NOT ONLY TO HIS OWN INTERESTS, BUT ALSO TO THE INTERESTS OF OTHERS. HAVE THIS MIND AMONG YOURSELVES, WHICH IS YOURS IN CHRIST JESUS, WHO, THOUGH HE WAS IN THE FORM OF GOD, DID NOT COUNT EQUALITY WITH GOD A THING TO BE GRASPED, BUT MADE HIMSELF NOTHING, TAKING THE FORM OF A SERVANT.

WHY WE ARE IN THIS BUSINESS

CHAPLAIN (LIEUTENANT COLONEL) DAVID CAUSEY, U.S. ARMY (RET)

THEREFORE, MY BELOVED BRETHREN, BE STEADFAST, IMMOVABLE, ALWAYS ABOUNDING IN THE WORK OF THE LORD, KNOWING THAT YOUR TOIL FOR THE LORD IS NEVER IN VAIN. *1 CORINTHIANS 15:58*

The United States Army Transport (USAT) Dorchester will always be remembered in the history of the Army Chaplain Corps. Upon this ship, the "Four Immortal Chaplains" (1LTs George Fox, Alexander Goode, Clark Poling, and John Washington) served. Together they saved the lives of Soldiers and Sailors after the doomed ship was torpedoed on February 3, 1943. They did everything to help load panic-stricken men into lifeboats. When there was nothing else, they gave away their coats, gloves, and life preservers to save a final few. Then they locked arm-in-arm and determined to encourage and pray for the dying and go down with the ship.

The testimony of their sacrifice has always served as the chaplains' "reality check" to get them back on course when they lose sight of their mission and higher calling. Yes, just mention the Dorchester, and the heroism of these four chaplains comes to mind.

But this story includes the name of another ship—the USS Escanaba. Both ships started as civilian vessels. After recruitment, the Escanaba transitioned into military service as a Coast Guard cutter.

Actually, it was the Escanaba that rescued most of the 230 survivors from the Dorchester's 904 passengers and crew. In fact, its rescue efforts broke new ground in lifesaving. It was one of the first ships to employ rescue swimmers who tied survivors (too cold to grasp a rope) to rescue lines that hoisted them from 34-degree (F) waters to the safety of the ship. This was not the first or last of Escanaba's lifesaving efforts.

Although its primary mission was to kill U-boats, it saved hundreds of lives and protected dozens of ships. Unfortunately, Escanaba had a brief military career. Tragedy struck the ship only four months after assisting the Dorchester, another victim of a German U-boat attack. It sank in less than three minutes.

No ship was there to help the crew of the Escanaba. The delayed rescue resulted in only two of its 103 crew surviving.

The story of the Escanaba reminds me of many people—people who have dedicated their lives to helping and healing others. Yet, when they find themselves in trouble and hurting, no one seems to be available to help them.

ABANDONMENT

Many times servicemembers feel this way. This is especially true for those who have sacrificed greatly to accomplish the military's mission. Then, when their lives are broken, the military seems to direct all its efforts to punishing rather than helping those servicemembers. They feel a profound sense of abandonment. *I was there for the military, but they weren't there for me.* Believe me. I sympathize with those who feel this way. I have experienced those same feelings and the fear of abandonment. But I need to remind myself that the outcome and rewards for the good, which I do in this present life will probably not return to me until the life to come.

The concept of *karma*—the idea that the good or bad that we do to others will come back to bless or curse us—is deep in our culture. But I've got to realize that this life is for service, service to God and to my fellow human beings. That's why we are in this business. This life is not the place for rewards. Heaven is the place where God will reward my faithfulness to Him and to others. In this life I must measure my success on how faithfully I use my days, hours, and resources for the advancement of God's Kingdom and for the betterment of others.

If a kindness I perform is someday returned to me, fine. But if not, I must always look to my heavenly Father as the One I serve and strive to please. I must look to Him as the One who will reward me in eternity. "Therefore, my beloved brethren, be steadfast, immovable, always abounding in the work of the Lord, knowing that your toil for the Lord is never in vain" (1 Corinthians 15:58).

DISCUSSION QUESTIONS

THE FOLLOWING DISCUSSION QUESTIONS ARE BASED ON THE SESSION OVERVIEW, THE ARTICLE, "WHY WE ARE IN THIS BUSINESS," AND KEY SCRIPTURES FOR THIS SESSION.

1. What does service mean to you? What factors or attitudes do you think must exist in order for someone, a warrior, to serve selflessly?

2. How does the story of the Four Chaplains or the efforts of the Escanaba inspire you to serve others? What particular Key Scriptures help you understand what's involved in this kind of service?

3. The author talks about "a profound sense of abandonment." What does he mean by this? Have you ever felt this way and what did you do about it? What helped? What hindered?

4. The article is titled "Why We Are In This Business." What is the point of this title? How does this point encourage you to be in the business of service?

5. How does this session help you reexamine your priorities and goals? What would you begin to do differently?

AFTER ACTION REVIEW

1. What new information did you learn about serving selflessly?

2. List the positive and negative attitudes that affect selfless service?

3. What does the author mean by abandonment? How do you plan to serve selflessly regardless?

4. How has this session helped you in your spiritual journey?

PERSONAL REFLECTION

1. Describe a time when you felt like you gave so much and received nothing in return. How has this session changed your perspective on selfless service?

2. What attitudes of service do you need to check in yourself? What attitudes do you need to incorporate?

3. How has this session helped you personally grow in your faith or your understanding of what God expects of you?

LET'S GO DEEPER.

REVIEW JOHN 13:5, 14 & PHILIPPIANS 2:3-7a.

1. What did Jesus mean by washing the disciples' feet?

2. What did He mean when He said, "You also ought to wash one another's feet"?

3. What are some practical ways you can look out for the interests of others?

4. What does the writer mean by "Have this mind among yourselves, which is yours in Christ Jesus"?

REVIEW LUKE 22:26-27 & MATTHEW 20:26-28

1. What is the secret of greatness?

2. How can you be a servant-leader?

NOTES

IN THIS LIFE I MUST MEASURE MY SUCCESS ON HOW FAITHFULLY I USE MY DAYS, HOURS, AND RESOURCES FOR THE ADVANCEMENT OF GOD'S KINGDOM AND FOR THE BETTERMENT OF OTHERS.

DAVID CAUSEY

SESSION TWELVE

Strength

SESSION OVERVIEW:

No one is ever strong enough to handle everything effectively. Resources become depleted. Challenges become too great. Courage gives way to fears. Even the best warriors grow weary or find themselves facing obstacles that are too big. That was certainly the case when the army of Israel found itself up against the command of the Philistines, and namely one huge warrior, Goliath. Israel felt powerless and helpless in the face of such opposition. Read 1 Samuel 17 and learn about the hero, David, who succeeded in defeating Goliath through his reliance on God's strength.

There are many words used in the Bible to describe strength. Whether you look to the Old Testament or in the New, while different words are used, their meanings generally convey empowerment, renewal, might, refuge, fierceness, effectual working, and enablement. And, when faced with just the challenges of living, we recognize that these are the things we need to succeed.

In this session, we'll focus on the meaning of strength and explore its spiritual applications. After all, it is God who gives us the strength to face each day—the power to the faint and might to the weak and worn.

GROWTH OBJECTIVES

- TO RECOGNIZE THE SOURCE OF STRENGTH.
- TO LEARN HOW TO ACCESS GOD'S STRENGTH.
- TO IDENTIFY WEAKNESSES THAT ARE GIVEN TO MORAL EROSION.
- TO ESTABLISH NEW GOALS THAT REPRESENT A LASTING LEGACY.

KEY SCRIPTURES

1 TIMOTHY 6:12

Fight the good fight of the faith. Take hold of the eternal life to which you were called and about which you made the good confession in the presence of many witnesses.

ISAIAH 40:29-31

He gives power to the faint, and to him who has no might he increases strength. Even youths shall faint and be weary, and young men shall fall exhausted, but they who wait for the LORD shall renew their strength; they shall mount up with wings like eagles; they shall run and not be weary; they shall walk and not faint.

2 KINGS 6:16

He said, 'Do not be afraid, for those who are with us are more than those who are with them.'

EPHESIANS 6:10

Finally, be strong in the Lord and the strength of his might.

PSALM 27:1b; 28:7; 46:1

The LORD is the stronghold of my life; of whom shall I be afraid?

The LORD is my strength and my shield; in him my heart trusts, and I am helped...

The LORD is our refuge and strength, a very present help in trouble.

2 SAMUEL 22:23-24; 29-30; 33-35, 40a

For all his rules were before me, and from his statues, I did not turn aside. I was blameless before him, and I kept myself from guilt.

For you are my lamp, O LORD, and my God lightens my darkness. For by you I can run against a troop, and by my God, I can leap over a wall.

This God is my strong refuge and has made my way blameless. He made my feet like the feet of a deer and set me secure on the heights. He trains my hands for war so that my arms can bend a bow of bronze.

For you equipped me with strength for the battle.

PROVERBS 18:10

The name of the LORD is a strong tower; the righteous man runs into it and is safe.

HEBREWS 11:32-34

And what more shall I say? For time would fail me to tell of Gideon, Barak, Samson, Jephthah, of David and Samuel and the prophets—who through faith conquered kingdoms, enforced justice, obtained promises, stopped the mouths of lions, quenched the power of fire, escaped the edge of the sword, were made strong out of weakness, became mighty in war, put foreign armies to flight.

ROMANS 8:36-37

As it is written, 'For your sake we are being killed all the day long; we are regarded as sheep to be slaughtered.' No, in all these things we are more than conquerors through him who loved us.

ISAIAH 40:29-31

HE GIVES POWER TO THE FAINT, AND TO HIM WHO HAS NO MIGHT HE INCREASES STRENGTH. EVEN YOUTHS SHALL FAINT AND BE WEARY, AND YOUNG MEN SHALL FALL EXHAUSTED, BUT THEY WHO WAIT FOR THE LORD SHALL RENEW THEIR STRENGTH; THEY SHALL MOUNT UP WITH WINGS LIKE EAGLES; THEY SHALL RUN AND NOT BE WEARY; THEY SHALL WALK AND NOT FAINT.

HOW TO DEFINE STRENGTH

CHAPLAIN (LIEUTENANT COLONEL) DAVID CAUSEY, U.S. ARMY (RET)

BE STRONG IN THE LORD AND HIS MIGHTY POWER. *EPHESIANS 6:10*

Haystack Rock is a 235-foot high basalt monolith that protrudes out of the ocean in the Oregon town of Cannon Beach. Haystack Rock was originally an intrusion of molten volcanic rock that was forced upward. Having been formed thousands of years ago, it was once connected to the shore. But now it stands alone, while all the surrounding bedrock has been worn away by the relentless pounding of waves and wind.

Why did it resist the effects of erosion, while the sandstone and limestone are all gone? It stands alone because of its composition and strength. Haystack rock is made of basalt, a tough volcanic rock—much harder than the sandstone and limestone that once surrounded it.

Note that geologists, interpretive signs, and publications do not praise the absence of the eroded sandstone and limestone. They highlight that which has resisted the effects of time and erosion—the basalt monolith that remains.

Geologists do not consider Haystack Rock as being stubborn, resistant, or out of step with the vanished geological features. They believe this basalt monolith is an emblem and durable monument of strength, an icon of the northern Oregon coast. Nor do they credit the easily eroded sandstone and limestone for being progressive or tolerant because they caved and gave ground. Geologists consider those sedimentary rocks to be weak, incapable of surviving the ravages of wind and waves.

Similarly, firm Christian believers should not view themselves as dimwitted, intolerant, or out of touch with the times. These things do not apply to those who maintain the moral high ground in society or refuse to give in to temptations or yield to persecutions. Believers who stand firm in their convictions display strength, not weakness. Believers who do not yield to the relentless waves of criticism and societal rejection display fortitude.

As the basalt monolith is wave-and wind-resistant, so the strong, believing Christian is temptation and sin-resistant. That kind of Christian stands firm against the winds of moral change. That person doesn't cave to temptation at first glance. Instead, he or she draws strength from the One who resisted all temptation, conquered sin and death, and in victory imparted life to all who put their trust in Him—Jesus Christ.

The Scripture says, "Be strong in the Lord and his mighty power. Put on all of God's armor so that you will be able to stand firm against all strategies of the devil. For we are not fighting against flesh-and-blood enemies; but against evil rulers and authorities of the unseen world, against mighty powers in this dark world, and evil spirits in the heavenly places. Therefore, put on every piece of God's armor so you will be able to resist the enemy in the time of evil. Then after the battle, you will still be standing firm" (Ephesians 6:10-13, New Living Translation).

DISCUSSION QUESTIONS

THE FOLLOWING DISCUSSION QUESTIONS ARE BASED ON THE SESSION OVERVIEW, THE ARTICLE, "HOW TO DEFINE STRENGTH," AND KEY SCRIPTURES FOR THIS SESSION.

1. Haystack Rock has resisted the effects of time and erosion. It remains because of its unique composition. Using this as an analogy, what are the elements that contribute to or compose strength of human character, moral resiliency, or substantial commitment?

2. According to Hebrews 11:32-34, men and women were able to accomplish some pretty amazing things. What is attributed to these achievements? In what areas does God test our faith? How might faith in God contribute to your personal and spiritual strength?

3. In the article, David suggests the eroded elements of limestone or sandstone as being insignificant. As a metaphor, they represent the shifting and unstable values of an ever-changing culture. In what ways have you embraced those insignificant values of culture, the values that will never hold up or matter in the end? What can you do to guard against incorporating those things into your life? What specific steps might you take?

4. Like Haystack Rock, where or when are you going to take a stand? What will be your legacy?

5. In the military, leaders talk about "building combat power." An operation requires sufficient strength and capability to succeed. In a spiritual sense, how might you build combat power? What does Ephesians 6:10-13 suggest building combat power and defeating the enemy?

AFTER ACTION REVIEW

1. What new information did you learn about developing strength?

2. Given what you've learned about spiritual strength, how do we access that strength? What capabilities do we need to face the challenges of living in an evil and sin-cursed world?

3. If faith is the means of experiencing the power of God, what can you do to develop your faith? What specific steps might you take?

4. How has this session helped you in your spiritual journey?

PERSONAL REFLECTION

1. What "eroding elements" do you need to discard in your life? What are the lasting things you need to sustain or add to your life?

2. How have you caved to temptation and the shifting, unstable values of the culture? What are those things and how can you set them aside?

3. Given your review of the Key Scriptures, what does God say about strength? What lessons of power might you glean from the Key Scriptures?

LET'S GO DEEPER

REVIEW ISAIAH 40:29-31.

1. Complete the verse by filling in the blanks:

   He gives ______________ to the faint, and to him who has no ______________ he increases __________________. Even youths shall faint and be weary, and young men shall fall exhausted; but they who ___________ for the LORD shall _____________ their strength; they shall mount up with wings like eagles; they shall run and not be weary; they shall walk and not _______________.

2. What do you think it means to wait for the LORD?

3. How has God strengthened you as you waited on Him?

REVIEW 1 TIMOTHY 6:12.

1. How can you "fight the good fight of the faith"?

2. What do you think the writer means by "Take hold of the eternal life to which you were called"?

LET'S GO DEEPER CONT'D:

REVIEW 2 SAMUEL 22:23-24; 29-30; 33-35, 40a.

1. What do you think was David's secret for experiencing God's strength in battle?

2. Identify the following seven things that David reports about God:

   My God ________________ my darkness.

   For by you I can ________ against a troop.

   And by my God I can ____________ over a wall.

   This God is my strong ______________.

   He has set me ________________ on the heights.

   He _________________ my hands for war.

   For you __________________ me with strength for the battle.

NOTES

ONE WHO GAINS STRENGTH BY OVERCOMING OBSTACLES POSSESSES THE ONLY STRENGTH WHICH CAN OVERCOME ADVERSITY.

ALBERT SCHWEITZER

A Call to Action

DR. DAMON FRIEDMAN

The enemy is conducting all-out warfare against every person in this world. He's launched the greatest campaign in the history of mankind. Today, more than ever, people feel isolated and insignificant. This issue is so alarming that a huge population will come to the point of committing suicide, using their God-given freedom to take their freedom. It's an epidemic plaguing the veteran community. There is a war between good and evil, and the enemy continues to gain ground. Evil is devouring men, dividing families, and destroying nations. What are we to do?

The God of the universe, who holds all power in the palm of His hand, presents a powerful message. In God's combat manual, the Bible, He says each person is so important, He knows them by name. God, who commands the heaven's armies, is calling people across the globe to fight against the enemy's tyranny. God is looking for one warrior whose heart is fully committed to Him so that He will exude His strength in and through that person to change the world for good (2 Chronicles 16:9).

God has empowered us to fight for what matters in life. He has provided weapons of warfare to establish your position, defend your perimeter, and destroy the enemy. Psalm 144:1 says, "Blessed be the Lord, my rock who trains my hands for war and my fingers for battle."

A warrior must recognize the fight between good and evil and be able to understand the art of warfare. Warriors must have the courage to live, fight, bleed, and die with the brother by their side — for there is no greater love than to lay down your life for your brother (John 15:13). We need warriors with impeccable character and faith who will fight for the most significant cause in human history: freedom.

To be a skilled warrior, you must have discipline, and train hard! When I was assigned to 2d Force Reconnaissance Company in the Marines, our team prepared to be shooting experts, especially during close quarter battle, which meant maneuvering through rooms within a building. During a high threat protective security detail course, I shot 10,000 rounds with my primary weapon (an M4), and another 10,000 rounds with my alternate weapon (a Glock 19). By the end of the week, our team moved like lightning with incredible speed and remarkable precision. We were a force to be reckoned with, and we were ready for any dangerous situation we might encounter. Our team had a tremendous amount of confidence in our weapons and our ability to execute. God wants each warrior to take the same approach with our spiritual training.

When we enlist in God's Army, He provides a unique gift to each warrior. I like to call it a superpower and I believe it's our responsibility to identify it, harness it, and use it to fight evil for the sake of the Kingdom. Very much like military soldiers who conduct the nation's bidding, we Kingdom warriors must conduct the Kingdom's bidding. This selfless service is a major sacrifice, and we must take this duty extremely seriously. The stakes are high. Time is limited, and souls are on the line.

The Warrior's Ethos is a code to live by, a philosophy that drives the decision-making process that leads us to train, fight, and win. We must honor it. Make sure to prepare appropriately. Plan and count the cost. Train hard and then go to war. "Fight the good fight of the faith" (1 Timothy 6:12). "Remember the Lord, who is great and awesome, and fight for your brothers, your sons, your daughters, your wives, and your homes" (Nehemiah 4:14). Fight for your faith. Fight for your family. Fight for your friends. Fight to the finish.

Godspeed,

F. Damon Friedman

Ethos Heraldry

SOF Missions is a military and veteran's service organization made up of veterans, health care professionals, contractors, group leaders, missionaries, volunteers and everyday people who are passionate about making a global impact for God's Kingdom. SOF Missions empowers warriors to find purpose, be resilient, and live well. We develop and provide customized treatment plans (psychological, physical, social and spiritual) for veterans through our national network of partners. Our goal is to help warriors know God, the Commander of the Universe, and recognize their mate purpose in this life.

ase visit SOFMissions.com to check out the resources we provide for military members, erans, and their families.

The Warrior's Journey is an interactive online resource for military members, veterans and their families. It presents the message of faith as a path for them to find wholeness in everyday life. In cooperation with other military partnerships, foundations, and non-profits, TWJ presents a wide range of resources that cover the unique issues that they encounter in service and at home. The vision of The Warrior's Journey is that warriors and their families will live in wholeness and be equipped to vigate the issues of life. The mission of The Warrior's Journey is to help warriors d their families discover and grow in their journey with God by providing contextual portunities to participate in an active faith community of believers. The strategy of e Warrior's Journey is to interact online with warriors and their families in finding oleness in daily life by searching, presenting, connecting, growing, and belonging.

ase visit TheWarriorsJourney.org to check out the resources we have available for military mbers, veterans, and their families.